上海大学出版社

2005年上海大学博士学位论文 40

U0358903

非等谱发展方程族的类孤子解

- 作 者：宁同科
- 专 业：计算数学
- 导 师：陈登远　张大军

Shanghai University Doctoral
Dissertation（2005）

Soliton-like Solutions for Nonisospectral Equation Hierarchies

Candidate: Ning Tongke

Supervisor: Chen Dengyuan
Zhang Dajun

Major: Computational Mathematics

Shanghai University Press
· **Shanghai** ·

上 海 大 学

　　本论文经答辩委员会全体成员审查，确认符合上海大学博士学位论文质量要求.

答辩委员会名单：

主任：
　　　　楼森岳　　上海交通大学

委员：
　　　　李志斌　　华东师范大学
　　　　秦成林　　上海大学
　　　　马和平　　上海大学
　　　　范恩贵　　复旦大学

导师： 陈登远

答辩日期： 2005. 3. 8

摘　要

本文利用反散射变换、Hirota 方法、Wronskian 技巧研究了非等谱发展方程族和等谱方程的 τ 方程族的精确解以及解的性质. 第二章从 Lax Pair 出发, 当谱参数按一定的规律随时间 t 变化时得到 KdV 系统的方程族、mKdV 系统的方程族、sine-Gordon 系统的方程族和 AKNS 系统的方程族, 同时由 AKNS 系统的方程族约化得到 mKdV 系统的方程族、sine-Gordon 系统的方程族、非线性 Schrödinger 系统的方程族. 相应的等谱方程族、非等谱方程族以及 τ 方程族是其特例. 第三章通过反散射变换方法得到 KdV 系统的方程族的 N-孤子解的精确表达式, 进而约化得到等谱 KdV 方程族、非等谱 KdV 方程族以及 τ 方程族的 N-孤子解. 第四章通过反散射变换的方法得到 AKNS 系统的方程族的 N-孤子解的精确表达式, 进而约化得到了 KdV 系统方程族、mKdV 系统方程族、sine-Gordon 系统方程族、非线性 Schrödinger 系统方程族的 N-孤子解的精确表达式. 第五章利用 Hirota 方法、Wronskian 技巧, 获得非等谱 sine-Gordon 方程、非等谱非线性 Schrödinger 方程、KdV 系统的 τ 方程、mKdV 系统的 τ 方程、sine-Gordon 系统的 τ 方程、非线性 Schrödinger 系统的 τ 方程的 N-孤子解, 并考察相应解的性质, 得到与等谱方程既有共同的特征又独有的性质. 第六章给出谱参数对时间的导数是谱参数的线性函数时非等谱 AKNS 方程族和等谱方程族之间的规范变换与转换算子.

关键词: 非等谱方程, τ 方程, 反散射变换, Hirota 方法,
Wronskian 技巧, 精确解

Abstract

In this paper, We obtain the solutions for nonisospectral evolution equation heirarchies and τ-sysmetry hierarchies by the inverse scattering transform (IST), Hirota method and Wronskian technique, we also investigate the properties of these solutions. In chapter 2, under the condition which spectral parameter k evolves according to time, we derive out equation hierarchies for KdV system, mKdV system, sine-Gordon system and AKNS system. Equation hierarchies for KdV system, mKdV system, sine-Gordon system and nonlinear Schrödinger system are reduced from equation hierharchy for AKNS system. Specially, isospectral hierarchies, nonisospectral hierarchies and related τ-sysmetry hierarchies are special cases of them. In chapter 3, by IST, we obtain N-soliton solution to equation hierarchy for KdV system, and reduce to N-soliton solution to isospectral KdV hierarchy, nonisospectral KdV hierarchy and τ-symetry hierarchy. In chapter 4, by IST, we obtain N-soliton solution to equation hierarchy for AKNS system, and reduce to N-soliton solution to equation hierarchies for KdV system, mKdV system, sine-Gordon system and nonlinear Schrödinger system. In chapter 5, we investigate the nonisospectral sine-Godorn equation, nonisospectral nonlinear Schrödinger equation by Hirota method and Wroskian technique, meantime we also consider τ-symetry equation for KdV, mKdV, sine-Gordon, nonlinear Schrödinger systems by Hirota method. The properties of these solutions are investigated. In chapter 6, we give gauge transform between nonisospectral AKNS hierarchy and isospectral AKNS hierarchy.

Keywords: Nonisospectral Equation Hierarchy, τ equations, Inverse Scattering transformaton, Hirota method, Wronskian technique, Exact solution

目　　录

第一章 前　言

§1.1　引言

在物理学的众多领域如流体、固体、基本粒子、等离子体、凝聚态、超导、激光、非线性光学等的研究中，都出现了一些孤子方程，这些方程最典型例子是 KdV 方程、sine-Gordon 方程、非线性Schrödinger 等方程，它们都具有孤子或类孤子解. 孤子在相互碰撞过程中保持自己的传播方向和能量不改变，使它具有类似于粒子和波动的许多特征，大量的研究表明孤子这一奇特现象在自然界具有普适性，因而存在广阔的应用前景. 对孤子方程的研究是数学物理领域的一个方兴未艾的课题，也是非线性科学的前沿课题.

§1.2　Lax 可积方程的求解

对于线性偏微分方程的求解已经有一套成熟的理论和方法，如D'Alembert 行波法、分离变量法、球面影像法和 Fourier 分析法等. 但对于非线性偏微分方程，叠加原理不再成立，求解变得十分困难. 在这一领域，人们进行了大量的工作，取得了一些成功的方法，例如反散射变换（IST）、Bäcklund 变换、Hirota 双线性导数方法、Wronskian 技巧、对称等一系列方法，其中 1967 年，Gardner，Green，Kruskal 和 Miura[1]（GGKM）在求解 KdV 方程的哥西（Cauchy）问题时所采用的反散射变换方法在非线性发展方程的求解中具有里程碑的地位. 我们知道，对于空气动力学研究中提

出的 Burgers 方程

$$u_t + uu_x - \nu u_{xx} = 0 \qquad (1.2.1)$$

可通过 Cole-Hopf 变换

$$u = -2\nu \frac{\phi_x}{\phi} \qquad (1.2.2)$$

化为线性热传导方程 $\phi_t - \nu\phi_{xx} = 0$，从而得到 Burguers 方程的精确解. 由于 KdV 方程与 Burgers 方程外形相似，很多学者尝试寻找类似于 Cole-Hopf 变换的函数变换，将 KdV 方程"线性化"，从而达到研究 KdV 方程的目的，然而这些努力都没有成功. 1967 年 Miura[2]等人发现 mKdV 方程

$$v_t - 6v^2 v_x + v_{xxx} = 0 \qquad (1.2.3)$$

和 KdV 方程

$$u_t + 6uu_x + u_{xxx} = 0 \qquad (1.2.4)$$

之间存在 Miura 变换

$$u = -v^2 - v_x \qquad (1.2.5)$$

如果将 Cole-Hopf 变换

$$v = \frac{\phi_x}{\phi} \qquad (1.2.6)$$

代入 Miura 变换(1.2.5)，则给出

$$\phi_{xx} + u\phi = 0 \qquad (1.2.7)$$

注意到 KdV 方程在 Galilean 变换 $u \to u - \lambda$, $t \to t$, $x \to x + 6\lambda t$ 下不变，这样线性方程(1.2.7)就成

$$\phi_{xx} + (u - \lambda)\phi = 0 \qquad (1.2.8)$$

这一方程在量子力学[3]中称为一维 Schrödinger 方程，其中 ϕ 是波函

数,u 是位势,λ 是谱参数. 但是与量子力学的 Schrödinger 方程不同的是,u 作为 KdV 方程的解,应依赖于时间. 即是说在(1.2.8)中,必须把时间考虑为参数. 这样一来,本征值 λ 及波函数 ϕ 也依赖于时间,因此可以假定波函数有时间发展式

$$\phi_t = A\phi + B\phi_x \qquad (1.2.9)$$

当取 $A = \gamma + u_x$,$B = 4\lambda + 2u$,式中 γ 是任意常数时,在 $\lambda_t = 0$ 条件下,容易推出线性问题(1.2.8)与(1.2.9)相容等价于 u 满足 KdV 方程(1.2.4). 这种将非线性方程等价于一对线性问题的相容性条件称为"Lax 可积".

GGKM 求解 KdV 方程 Cauchy 问题

$$\begin{cases} u_t + 6uu_x + u_{xxx} = 0 \\ u(x, 0) = f(x) \end{cases}$$

可归纳为两大步,第一步是**正散射问题**,正散射问题是给定位势 u,求线性问题(1.2.8)的离散谱 λ_n、连续谱 λ 及其他散射数据并研究特征函数的性质.

根据量子力学散射理论,当位势 u 在无穷远处随 x 充分快地趋于零时,Schrödinger 谱问题(1.2.8)有特征函数平方可积的有限个离散谱 $\lambda = \kappa_n^2 > 0$,$n = 1, 2, \cdots, N$,以及特征函数有界的连续谱 $\lambda = -k^2 < 0$,其特征函数具有渐进式:

$$\text{对于离散谱 } \lambda = \kappa_n^2 > 0,\ x \to \infty \qquad (1.2.10)$$

$$\phi_n(x, t) \sim c_n(t)\exp(-\kappa_n x),\ \int_{-\infty}^{\infty} \phi_n^2 \,\mathrm{d}x = 1$$

对于连续谱 $\lambda = -k^2 < 0$,

(1) $x \to \infty$,$\phi(x, t) \sim \exp(-ikx) + R(k, t)\exp(ikx)$

$$(1.2.11a)$$

(2) $x \to -\infty$,$\phi(x, t) \sim T(k, t)\exp(-ikx)$ $\qquad (1.2.11b)$

这里(1.2.10)的 $c_n(t)$ 使得特征函数的平方在实轴上的积分等于 1,称其为离散特征函数的"归一化系数".(1.2.11a)第一项相应于波振幅为 1,几率密度为 k 的入射波(见图 1);第二项相应于经位势 $u(x)$ 反射后的反射波,其几率密度为 kR^2,两者之比为 $kR^2/k = R^2$,因此称 $R(k, t)$ 为"反射系数".(1.2.11b)相应于穿透位势 $u(x)$ 后继续前进的平面波,其几率密度为 kT^2,类似地称 $T(k, t)$ 为"透射系数".

集合

$$S(\lambda, t) = \{\{\kappa_n, c_n(t)\}_1^N, T(k, t), R(k, t)\} \quad (1.2.12)$$

就是量子力学的 Schrödinger 谱问题(1.2.8)的散射数据.

当 $t = 0$,对于给定的 $u(x, 0)$,由谱问题(1.2.8)得到散射数据

$$S(\lambda, 0) = \{\{\kappa_n, c_n(0)\}_1^N, T(k, 0), R(k, 0)\} \quad (1.2.13)$$

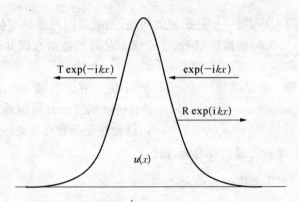

$T\exp(-\mathrm{i}kx)$ $\exp(-\mathrm{i}kx)$

$R\exp(\mathrm{i}kx)$

$u(x)$

图 1　入射波 $\mathrm{e}^{-\mathrm{i}kx}$ 遇到位势 u 发生反射与透射

第二步是**反问题**,反问题就是根据散射数据 $S(\lambda, t)$ 重构位势 $u(t, x)$.

在反问题这一阶段,分为两步:一是研究散射数据的时间演化规律,这一步可利用波函数的时间发展式(1.2.9),给出 t 时刻的散射数据(1.2.12)和 $t = 0$ 时刻的散射数据(1.2.13)之间的关系

$$\kappa_n = \mathrm{con}\,s\tan t,\ c_n(t) = c_n(0)\exp(4\kappa_n^3 t),\ n = 1,\,2,\,\cdots,\,N$$

$$(1.2.14a,\ b)$$

$$T(k,\,t) = T(k,\,0),\ R(k,\,t) = R(k,\,0)\exp(8\mathrm{i}k^3 t)$$

$$(1.2.14c,\ d)$$

第二步,也是关键一步就是位势重构. 1955 年, Gel'fand 和 Levitan[4]首先利用散射数据定义函数

$$F(x,\,t) = \sum_{n=1}^{N} c_n^2(t)\exp(-\kappa_n x) + \frac{1}{2\pi}\int_{-\infty}^{\infty} R(k,\,t)\exp(\mathrm{i}kx)\mathrm{d}k$$

$$(1.2.15)$$

然后求解线性积分方程,即 Gel'fand-Levitan-Marchenko 方程

$$K(x,\,y,\,t) + F(x+y,\,t) + \int_x^{\infty} K(x,\,z,\,t)F(z+y,\,t)\mathrm{d}z = 0,\ (y > 0)$$

$$(1.2.16)$$

得到 $K(x,\,y,\,t)$,则位势 $u(x,\,t)$ 可以由下式恢复

$$u(x,\,t) = 2\frac{\mathrm{d}K(x,\,x,\,t)}{\mathrm{d}x} \qquad (1.2.17)$$

这种方法类似于解线性方程的 Fourier 变换方法,因此也被称为"非线性 Fourier 变换". IST 求解的步骤可以用图 2 来概括.

图 2　IST 求解的步骤

1968 年 Lax[5]将 GGKM 上述思想放到更一般的框架上,为 IST 作为求解非线性发展方程的一种方法铺平了道路. 即:对于任意

非线性发展方程

$$u_t = K(u), \; K \text{ 是非线性算子} \quad (1.2.18)$$

如果存在依赖于 u 的线性算子 L 和 M，使得非线性发展方程（1.2.18）成为 线性问题

$$L\phi = \lambda\phi \quad (1.2.19a)$$

$$\phi_t = M\phi \quad (1.2.19b)$$

的相容性条件

$$L_t = [M, L] + \lambda_t \quad (1.2.20)$$

则称线性问题（1.2.9）为非线性发展方程（1.2.18）的 Lax Pair，（1.2.20）称为非线性发展方程的 Lax 表示，具有这种性质的非线性发展方程被称为是 Lax 可积. 则对这种非线性发展方程就可以应用 IST 法求解.

1973 年 Ablowitz[6，7]等人在前人工作的基础上，考察了 2×2 阶矩阵谱问题，得到了一大类物理上感兴趣的非线性发展方程，成功地给出了反散射解，并将这种求解方法命名为"反散射变换-非线性发展方程的 Fourier 变换". 之后 Ablowitz 和 Haberman[8]又将 2×2 阶矩阵谱问题推广到了 $n \times n$ 阶矩阵谱问题，并且非线性发展方程所含的空间变量可以是高维的，进一步扩大了反散射变换的求解范围，诸如 Boussinesq 方程[8，9]、非线性 Schrödinger 方程[9]、mKdV 方程[10]、sine-Gordon 方程[11]等，并成功地推广到高维系统[12，13]和微分-差分方程[13，14]，相当一大类非线性发展方程都有反散射解[15]. 在国内李翀神、庄大蔚[16]证明了 Kaup-Newell 谱问题和 AKNS 谱问题反散射变换的散射量和平移变换的等价性；顾新身[17]将反散射变换应用到具有二重根特征根的复 KdV 方程，得到了解；曾云波等[18，19]推广反散射变换到自容源方程获得具有非定常速度的孤子波. 最近，陈登远、宁同科和张大军将反散射变换应用到非等谱方程族[20，21]得到了整个方程族的解. 反散射变换作为求解非线性发展方程的方法仍在不断地发展中.

反散射变换自 1976 年由 GGKM 等人提出以来,取得了相当大的发展,所适用的方程类也在不断扩大,但这一方法要成为一种系统完整的方法还有很长的路要走,有许多问题仍有待解决,例如:

1. 什么样的非线性发展方程可以作为一组线性问题的可积条件?什么样的非线性发展方程可以用 IST 求解?

2. IST 将一个非线性问题转化为线性问题,但对于这些线性问题本身尚缺乏有效的处理方法.比如 IST 求得的解,都是在假设反射系数为零的条件下得到的纯孤子解,如何寻找反射系数不为零又有离散谱的解?

3. 目前已有的 IST 可解的方程,大多要求相应的线性问题的特征值 λ 与时间 t 无关,这一要求能否放弃?

4. 如何利用 IST 解非线性发展方程的混合问题?

5. 如何利用 IST 得到 Hirota 形式丰富多样的解?

这些都是非常重要而有趣的问题.

Bäcklund 变换及 Darboux 变换[22-25]也是寻求孤子方程解的重要方法,孤子方程通常存在将所对应的线性问题化为自身的规范变换,即 Darboux 变换.这时孤子方程两个解之间满足一定的关系式,这种解之间的关系即为孤子方程的 Bäcklund 变换.Bäcklund 变换将求解高阶微分方程转化为求包含解之间关系的较低阶的微分方程组.利用 Bäcklund 变换,可以从已知解出发,求出新的孤子解.但该方法涉及解微分方程组,往往在求多孤子解时遇到麻烦.直到 1974 年,Hirota[26]给出了一种 Bäcklund 变换的双线性导数形式,使得求多孤子解变得简单起来.Bäcklund 变换的双线性形式可以从双线性方程得到[26],也可以从普通形式的 Bäcklund 变换或者 Darboux 变换得到[27],不同的形式具有等价性[27].

Hirota 双线性导数方法[28]是求解一大类非线性发展方程的有力工具.它的求解过程可以归纳为:(1)引入因变量 u 的变换,将原方程改写成双线性导数形式;(2)从双线性方程出发构造 N-孤子解.该方法以双线性导数为工具,与非线性方程的线性问题(Lax

Pair)无关,操作简便,可以周而复始地进行下去,方便地给出 N-孤子解,其适用范围涵盖了所有反散射变换可解的方程([29,30-35]);而且成功地推广到了离散孤子系统 [36-42]. 这一方法和反散射变换的区别之处是只能求解单个方程而无法处理整个方程族,而且其对 N-孤子解表达式的猜测难以给出令人满意的证明. 另一方面,Hirota 方法引入的因变量变换,是在反散射解的启发下给出的,对于一个尚没有反散射解的方程,如何引入因变量变换将原方程化为双线性方程需要很高的技巧,没有一定的方法可循,这就使得它的应用受到了很大的限制.

Wronskian 技巧是求解孤子方程的又一直接方法. 它首先由 Satsuma[43] 在 1979 年引入,而作为求解孤子方程的一种方法,则是由 Freeman 和 Nimmo[44] 在 1983 年首先完善的. 该方法以 Hirota 方法为基础,即首先要得到孤子方程的双线性形式或双线性 Bäcklund 变换,然后适当选择 ϕ_j 构成 Wronskian 行列式 $W(\phi_1, \phi_2, \cdots, \phi_N)$. Wronskian 行列式由于其良好的性质,便于直接代到双线性方程或双线性 Bäcklund 变换中进行验证,因此被广泛应用. Freeman 等人应用该方法成功地获得了一系列方程的 Wronskian 形式的解:KdV 方程和 KP 方程[44],Boussinesq 方程[45],球 Boussinessq 方程[46],非局部 Boussinesq 方程[47],mKdV 方程[48],sine-Gordon 方程[48],非线性 Schrödinger 方程[49,50],Davey-Stewartson 方程[50],Toda 链[51],二维 Toda 链等等[52-55].

当然,求孤子方程的精确解的方法远不止于此,并且不断有新的方法出现. 比如,最近,韩文亭和李翊神提出了一种构造孤子方程解的矩阵方法[56,57];Hirota[58] 通过矩阵的 Pfaffian 表示得到了 BKP 方程的解;曾云波等提出了通过约束流来构造 N-孤子解的方法[59,60];陈登远、张大军和邓淑芳[61] 利用 Hirota 方法构造出了孤子方程的新解,其中部分方程的解具有奇异性. 齐次平衡法[62,63]、混合指数法[64]、双曲正切法[65]、Jacobi 椭圆函数展开法[66]等等,都是构造一些非线性发展方程解析解的有效方法. 通过这些方法所获得的非线性发展方程的诸多精确解,合理地解释了一些自然现象,极大地推动了相关学科如物理学、力学、应用数学以及工程技术的发展.

§1.3 本文的选题和主要工作

正如前面所述,反散射变换法在孤子方程求解中具有非常重要的地位,已有的结果大多是针对等谱方程族,少数文献讨论了谱参数对时间的导数是谱参数的线性函数的情形[67,68],或者通过 Darboux 变换[69,70]讨论了整个非等谱方程族的求解,但没能给出解的具体表达式.另一方面,我们知道,联系于同一谱问题的等谱流 K_l 和非等谱流 σ_l[71,72]在交换子运算 $[\![\cdot,\cdot]\!]$ 下构成 Lie 代数 [73-78]:

$$[\![K_l, K_j]\!] = 0 \qquad (1.3.1a)$$

$$[\![K_l, \sigma_j]\!] = (al+b)K_{l+j+c} \qquad (1.3.1b)$$

$$[\![\sigma_l, \sigma_j]\!] = a(l-j)\sigma_{l+j+c} \qquad (1.3.1c)$$

由此可以得到一般的等谱发展方程族

$$u_t = K_l \qquad (1.3.2)$$

具有两组对称 K-对称和 τ-对称 $\tau_l = t(aj+b)K_{j+l+c} + \sigma_l$,这两组对称构成 Lie 代数,即:

$$[\![K_l, K_j]\!] = 0 \qquad (1.3.3a)$$

$$[\![K_l, \tau_j]\!] = (al+b)K_{l+j+c} \qquad (1.3.3b)$$

$$[\![\tau_l, \tau_j]\!] = a(l-j)\tau_{l+j+c} \qquad (1.3.3c)$$

由 K-对称构成的方程正好是等谱方程族,由 τ-对称构成的方程族也是非等谱方程族的一种,求解非等谱方程族和 τ-方程族正是本文的任务.

无论是非等谱方程族还是 τ-方程族,都和等谱方程族具有相同形式的谱问题,因而利用 IST 求解时,正散射问题是相同的,困难在于对非等谱方程族和 τ-方程族如何确定散射数据随时间的演化关系.

本文的主要工作如下:

1. 给出从 Lax pair 得到 KdV、mKdV、sine-Gordon 系统的方程族和 AKNS 系统的方程族,以及由 AKNS 系统的方程族约化得到 mKdV 系统的方程族,非线性 Schrödinger 系统的方程族,sine-Gordon 系统的方程族.

2. 利用反散射变换分别得到 KdV 系统的方程族,AKNS 系统的方程族的精确解表达式,并约化得到 KdV 系统的方程族,mKdV 系统的方程族,非线性 Schrödinger 系统的方程族,sine-Gordon 系统的方程族的解. 而等谱方程族、非等谱方程族与 τ 方程族的解是其特殊情形.

3. 利用 Hirota 方法、Wronskian 技巧分别讨论一类非等谱方程的解,前者与反散射变换的解相一致,后者与反散射变换的解有区别,并研究解的性质.

4. 导出当谱参数关于时间的导数是谱参数的线性函数时的非等谱 AKNS 方程族和等谱 AKNS 方程族之间相应的规范变换与转换算子.

第二章 一类 Lax 可积的非线性 发展方程族的导出

§2.1 KdV 系统的方程族的导出

本节从 Schrödinger 谱问题出发,在适当的条件下,得到 KdV 系统方程族. 等谱 KdV 方程族、非等谱 KdV 方程族以及等谱 KdV 方程族的 τ 方程族都可以作为它的特例.

考虑 Schrödinger 谱问题

$$\phi_{xx} + u\phi = \lambda\phi \qquad (2.1.1a)$$

和时间发展式

$$\phi_t = A\phi + B\phi_x \qquad (2.1.1b)$$

其中 λ 是谱参数,满足 $\lambda_t = \dfrac{\beta(t)}{2}(4\lambda)^{n+1}$, A, B 是位势 u 和谱参数 λ 的待定函数,且满足边值条件

$$A\big|_{u=0} = -\frac{\beta(t)}{2}(4\lambda)^n \qquad (2.1.2a)$$

$$B\big|_{u=0} = \alpha(t)(4\lambda)^{n+s-1} + \beta(t)(4\lambda)^n x \qquad (2.1.2b)$$

由谱问题(2.1.1)的相容性条件 $\phi_{xxt} = \phi_{txx}$,给出

$$(2A_x + B_{xx})\phi_x + \left[u_t - \frac{\beta(t)}{2}(4\lambda)^{n+1} + A_{xx} + 2(\lambda - u)B_x - u_x B\right]\phi = 0$$

$$(2.1.3)$$

于是推知

$$2A_x + B_{xx} = 0 \qquad (2.1.4a)$$

$$u_t = -A_{xx} - 2(\lambda - u)B_x + u_x B + \frac{\beta(t)}{2}(4\lambda)^{n+1} \quad (2.1.4b)$$

从(2.1.4)中消去 A，得

$$u_t = 2\left(\frac{1}{4}\partial^3 + u\partial + \frac{1}{2}u_x\right)B - 2\lambda B_x + \frac{\beta(t)}{2}(4\lambda)^{n+1}$$

$$= \frac{1}{2}TB_x - 2\lambda B_x + \frac{\beta(t)}{2}(4\lambda)^{n+1} \qquad (2.1.5)$$

其中 $T = \partial^2 + 4u + 2u_x\partial^{-1}$ 是 KdV 递推算子.

设 B 可按 λ 展成 $n+s-1$ 次多项式

$$B = \sum_{j=0}^{n+s-1} b_j \lambda^{n+s-1-j} \qquad (2.1.6)$$

将其代入到(2.1.5)中，并令 λ 的同次幂系数相等得到如下关系式

$$u_t = \frac{1}{2}Tb_{n+s-1, x} \qquad (2.1.7a)$$

$$b_{j, x} = \frac{1}{4}Tb_{j-1, x}, \ (j = 1, 2, \cdots, s-2, s, s+1, \cdots, n+s-1)$$

$$(2.1.7b)$$

$$b_{s-1, x} = \frac{1}{4}Tb_{s-2, x} + 4^n\beta(t) \qquad (2.1.7c)$$

$$b_{0, x} = 0 \qquad (2.1.7d)$$

根据 B 所满足的边值条件(2.1.2b)，有

$$b_0 = 4^{n+s-1}\alpha(t) \qquad (2.1.8)$$

于是由(2.1.7b, c)和(2.1.8)递推算得

$$b_{j,x} = \left(\frac{1}{4}T\right)^j b_{0,x}, \quad (j=1, 2, \cdots, s-2) \qquad (2.1.9a)$$

$$b_{j,x} = \left(\frac{1}{4}T\right)^j b_{0,x} + 4^n \left(\frac{1}{4}T\right)^{j-s+1}\beta(t), \quad (j=s-1, s, \cdots, n+s-1)$$

$$(2.1.9b)$$

将(2.1.9b)代入(2.1.7a)给出

$$u_t = \alpha(t)T^{n+s-1}u_x + \beta(t)T^n(xu_x+2u) \qquad (2.1.10)$$

如果 $\beta(t)=0$,这时谱参数不随时间而改变,(2.1.10)化为等谱方程族

$$\dot{u_t} = \alpha(t)T^{n+s-1}u_x \qquad (2.1.11a)$$

如果 $\alpha(t)=0$,则得非等谱方程族

$$u_t = \beta(t)T^n(xu_x+2u) \qquad (2.1.11b)$$

因此,我们称 $K_n = T^n u_x$ 为 n 阶 KdV 等谱流,$\sigma_n = T^n(xu_x+2u)$ 为 n 阶 KdV 非等谱流.

定理 2.1 设 $\alpha(t)$,$\beta(t)$ 是关于时间 t 的任意函数,如果谱参数 λ 随时间的变化规律为

$$\lambda_t = \frac{\beta(t)}{2}(4\lambda)^{n+1}$$

待定函数 A, B 满足边值条件(2.1.2),则有

$$B = x(4\lambda)^n\beta(t) + \alpha(t)(4\lambda)^{n+s-1} + 2\alpha(t)\sum_{j=1}^{n+s-1}(4\lambda)^{n+s-1-j}\partial^{-1}T^{j-1}u_x +$$

$$2\beta(t)\sum_{j=s}^{n+s-1}(4\lambda)^{n+s-1-j}\partial^{-1}T^{s-j}(xu_x+2u) \qquad (2.1.12)$$

和 KdV 系统的方程族

$$u_t = \alpha(t)T^{n+s-1}u_x + \beta(t)T^n(xu_x+2u)$$

特别地,(1) 当 $\alpha(t) = 1$,$\beta(t) = 0$,有

$$u_t = T^{n+s-1} u_x,\ n+s = 1,\ 2,\ \cdots \qquad (2.1.13a)$$

即为等谱 KdV 方程族[27,79,80],其前两个非平凡的方程为

$$u_t = K_1 \equiv u_{xxx} + 6uu_x \qquad (2.1.13b)$$

$$u_t = K_2 \equiv u_{xxxxx} + 10uu_{xxx} + 20u_x u_{xx} + 30u^2 u_x$$
$$(2.1.13c)$$

(2) 当 $\alpha(t) = 0$,$\beta(t) = 1$,有

$$u_t = T^n (xu_x + 2u) \qquad (2.1.14a)$$

即为非等谱 KdV 方程族[27,71,80];其前两个非平凡的方程为

$$u_t = xK_1 + 4u_{xx} + 8u^2 + 2u_x \partial^{-1} u \qquad (2.1.14b)$$

$$u_t = xK_2 + 6K_{1,x} + 12uu_{xx} + 32u^3 + 2K_1 \partial^{-1} u + 6u_x \partial^{-1} u^2$$
$$(2.1.14c)$$

(3) 当 $\alpha(t) = (2s-1)t$,$\beta(t) = 1$,有

$$u_t = (2s-1)tT^{n+s-1} u_x + T^n (xu_x + 2u) \qquad (2.1.15a)$$

即为 KdV 系统的 τ 方程族[77].

当 $s = 2$,$n = 0$,τ 方程为

$$u_t = 3t(u_{xxx} + 6uu_x) + xu_x + 2u \qquad (2.1.15b)$$

§2.2 mKdV 系统的方程族和 sine-Gordon 系统的方程族的导出

§2.2.1 mKdV 系统的方程族的导出

考虑另一个 Schrödinger 谱问题

$$\psi_{xx} + 2v\psi_x = \lambda\psi \qquad (2.2.1a)$$

与时间发展式

$$\psi_t = A\psi + B\psi_x \qquad (2.2.1b)$$

其中 λ 是谱参数,满足 $\lambda_t = \dfrac{\beta(t)}{2}(4\lambda)^{n+1}$, A, B 是位势 v 和谱参数 λ 的待定函数,且满足边值条件

$$A\big|_{v=0} = 0, \; B\big|_{v=0} = \alpha(t)(4\lambda)^{n+s} + \beta(t)(4\lambda)^n x \qquad (2.2.2a)$$

由(2.2.1)的相容性条件 $\psi_{xxt} = \psi_{txx}$,得

$$\left[A_{xx} + 2vA_x + 2\lambda B_x - \frac{\beta(t)}{2}(4\lambda)^{n+1} \right]\psi +$$

$$(2v_t + B_{xx} - 2vB_x - 2v_xB + 2A_x)\psi_x = 0 \qquad (2.2.3)$$

令

$$A_{xx} + 2vA_x + 2\lambda B_x - \frac{\beta(t)}{2}(4\lambda)^{n+1} = 0 \qquad (2.2.4)$$

则有

$$v_t = -\frac{1}{2}B_{xx} + vB_x + v_xB - A_x \qquad (2.2.5)$$

由(2.2.4)可以给出

$$B = -\frac{1}{2\lambda}\left[A_x + 2\partial^{-1}vA_x - x\frac{\beta(t)}{2}(4\lambda)^{n+1} \right] + B_0 \qquad (2.2.6a)$$

其中 B_0 是积分常数.

利用 A 和 B 所满足的边值条件,容易得到 $B_0 = \alpha(t)(4\lambda)^{n+s}$,因此

$$B = -\frac{1}{2\lambda}\left[A_x + 2\partial^{-1}vA_x - x\frac{\beta(t)}{2}(4\lambda)^{n+1} \right] + \alpha(t)(4\lambda)^{n+s}$$

$$(2.2.6b)$$

将(2.2.6b)代入(2.2.5)，有

$$\lambda v_t = \left(\frac{1}{4}\partial^2 - v^2 - v_x\partial^{-1}v\right)A_x + 4^n\beta(t)(xv)_x\lambda^{n+1} - \lambda A_x + 4^{n+s}\alpha(t)v_x\lambda^{n+s+1}$$

$$= \frac{1}{4}FA_x - \lambda A_x + 4^n\beta(t)(xv)_x\lambda^{n+1} + 4^{n+s}\alpha(t)v_x\lambda^{n+s+1}$$

$$(2.2.7)$$

式中 $F = \partial^2 - 4v^2 - 4v_x\partial^{-1}v$ 是 mKdV 递推算子.

设 A 是 λ 的 $n+s$ 次多项式

$$A = \sum_{j=1}^{n+s} a_j\lambda^{n+s+1-j} \qquad (2.2.8)$$

将其代入(2.2.7)并比较 λ 的同次幂系数,得

$$v_t = \frac{1}{4}Fa_{n+s,x} \qquad (2.2.9a)$$

$$a_{s+1,x} = \frac{1}{4}Fa_{s,x} + 4^n\beta(t)(xv)_x \qquad (2.2.9b)$$

$$a_{j,x} = \frac{1}{4}Fa_{j-1,x}, \ (j=1,2,\cdots,s,s+2,\cdots,n+s)$$

$$(2.2.9c)$$

$$a_{1,x} = 4^{n+s}\alpha(t)v_x \qquad (2.2.9d)$$

由(2.2.9c)和(2.2.9b)可以归纳出

$$a_{j,x} = \left(\frac{1}{4}F\right)^j a_{1,x}, \ (j=2,\cdots,s) \qquad (2.2.10a)$$

$$a_{j,x} = \left(\frac{1}{4}F\right)^j a_{1,x} + 4^n\beta(t)\left(\frac{1}{4}F\right)^{j-s-1}(xv)_x, \ (j=s+1,\cdots,n+s)$$

$$(2.2.10b)$$

将(2.2.10b)代入(2.2.9a)就得到 mKdV 系统的方程族

$$v_t = \alpha(t)F^{n+s}v_x + \beta(t)F^n(xv)_x \qquad (2.2.11)$$

如果 $\beta(t) = 0$,这时谱参数不随时间改变,(2.2.11)化为等谱方程族

$$v_t = \alpha(t)F^{n+s}v_x \qquad (2.2.12a)$$

如果 $\alpha(t) = 0$,则有非等谱方程族

$$v_t = \beta(t)F^n(xv)_x \qquad (2.2.12b)$$

因此,我们称 $K'_n = F^n v_x$ 为 n 阶 mKdV 等谱流,$\sigma'_n = F^n(xv)_x$ 为 n 阶 mKdV 非等谱流.

定理 2.2 设 $\alpha(t)$,$\beta(t)$ 是关于时间 t 的任意函数,如果谱参数 λ 随时间的变化规律为

$$\lambda_t = \frac{1}{2}\beta(t)(4\lambda)^{n+1}$$

待定函数 A,B 满足边值条件(2.2.2),则有 mKdV 系统的方程族

$$v_t = \alpha(t)F^{n+s}v_x + \beta(t)F^n(xv)_x$$

特别地,(1) 当 $\alpha(t) = 1$,$\beta(t) = 0$,有

$$v_t = F^{n+s}v_x \qquad (2.2.13a)$$

即为等谱 mKdV 方程族[27,79,80];其前两个非平凡的方程为

$$v_t = K'_1 \equiv v_{xxx} - 6v^2 v_x \qquad (2.2.13b)$$

$$v_t = K'_2 \equiv v_{xxxxx} - 10v^2 v_{xxx} - 40vv_x v_{xx} - 10v_x^3 + 30v^4 v_x$$
$$\qquad (2.2.13c)$$

(2) 当 $\alpha(t) = 0$,$\beta(t) = 1$,有

$$v_t = F^n(xv)_x \qquad (2.2.14a)$$

即为非等谱 mKdV 方程族[27,71,80];其前两个非平凡的方程为

$$v_t = xK'_1 + 3v_{xx} - 4v^3 - 2v_x\partial^{-1}v^2 \qquad (2.2.14\text{b})$$

$$v_t = xK'_2 + 5K'_{1,x} - 10v^2 v_{xx} + 16v^5 - 2K_1\partial^{-1}v^2 + 6v_x\partial^{-1}v_x^2 + 6v_x\partial^{-1}v^4$$

$$(2.2.14\text{c})$$

(3) 当 $\alpha(t) = (2s+1)t$，$\beta(t) = 1$，有

$$v_t = (2s+1)tF^{n+s}v_x + F^n(xv)_x \qquad (2.2.15\text{a})$$

即为 mKdV 系统的 τ 对称方程族[77].

当 $n = 0$，$s = 1$ 时，τ 方程为

$$v_t = 3t(v_{xxx} - 6v^2 v_x) + (xv)_x \qquad (2.2.15\text{b})$$

§2.2.2　sine-Gordon 系统的方程族的导出

在(2.2.1)中，以 iv_x 代替 $2v$，以 $-s$ 代 s，$-n$ 代 n，而其他条件不变，即：谱参数 λ 随时间 t 的变化规律满足 $\lambda_t = \dfrac{\beta(t)}{2}(4\lambda)^{1-n}$，和 A，B 满足边值条件 $A\big|_{u=0} = 0$，$B\big|_{u=0} = \alpha(t)(4\lambda)^{-n-s} + \beta(t)(4\lambda)^{-n}x$ 时，相容性条件(2.2.3)变为

$$\left[A_{xx} + iv_x A_x + 2\lambda B_x - \frac{\beta(t)}{2}(4\lambda)^{1-n}\right]\psi +$$

$$(iv_{xt} + B_{xx} - iv_x B_x - iv_{xx}B + 2A_x)\psi_x = 0 \qquad (2.2.16)$$

令

$$A_{xx} + iv_x A_x + 2\lambda B_x - \frac{\beta(t)}{2}(4\lambda)^{1-n} = 0 \qquad (2.2.17)$$

则有

$$iv_{xt} = -B_{xx} + iv_x B_x + iv_{xx}B - 2A_x \qquad (2.2.18)$$

由(2.2.17)并利用 B 所满足的边值条件，得

$$B = -\frac{1}{2\lambda}\left[A_x + \mathrm{i}\partial^{-1}v_x A_x - x\frac{\beta(t)}{2}(4\lambda)^{1-n}\right] + \alpha(t)(4\lambda)^{-n-s}$$

$$(2.2.19)$$

代入(2.2.18)，有

$$\lambda v_{xt} = -\frac{\mathrm{i}}{2}A_{xxx} - \frac{\mathrm{i}}{2}(v_x\partial^{-1}v_x A_x)_x + 2\mathrm{i}\lambda A_x +$$

$$\frac{\beta(t)}{4}(4\lambda)^{1-n}(xv_x)_x + \frac{\alpha(t)}{4}(4\lambda)^{1-n-s}v_{xx} \quad (2.2.20)$$

令 $\mathrm{i}\dfrac{T_x}{v_x} = A_x$，则上式可以改写为

$$\lambda v_{xt} = \frac{1}{2}\left(\frac{T_x}{v_x}\right)_{xx} + \frac{1}{2}(v_x T)_x - 2\lambda\frac{T_x}{v_x} + \frac{\beta(t)}{4}(4\lambda)^{1-n}(xv_x)_x +$$

$$\frac{\alpha(t)}{4}(4\lambda)^{1-n-s}v_{xx} \quad (2.2.21)$$

将 T 展开为 λ 的负幂多项式

$$T = \sum_{j=1}^{n+s}T_j\lambda^{-n-s+j} \quad (2.2.22)$$

将(2.2.22)代入(2.2.21)比较 λ 的同次幂系数给出

$$v_{xt} = -2\frac{T_{n+s-1,\,x}}{v_x} \quad (2.2.23a)$$

$$\frac{1}{2}\left(\frac{T_{1,\,x}}{v_x}\right)_{xx} + \frac{1}{2}(v_x T_1)_x + 4^{-n-s}\alpha(t)(xv_x)_x = 0$$

$$(2.2.23b)$$

$$\frac{1}{2}\left(\frac{T_{j,\,x}}{v_x}\right)_{xx} + \frac{1}{2}(v_x T_j)_x - 2\frac{T_{j-1,\,x}}{v_x} = 0, \quad (j = 1, 2, \cdots, s)$$

$$(2.2.23c)$$

$$\frac{1}{2}\left(\frac{T_{s+1,\,x}}{v_x}\right)_{xx} + \frac{1}{2}(v_x T_{s+1})_x - 2\frac{T_{s,\,x}}{v_x} + 4^{-n}\beta(t)(xv_x)_x = 0$$

$$(2.2.23d)$$

$$\frac{1}{2}\left(\frac{T_{j,\,x}}{v_x}\right)_{xx} + \frac{1}{2}(v_x T_j)_x - 2\frac{T_{j-1,\,x}}{v_x} = 0,\ (j = s+2, \cdots, n+s)$$

$$(2.2.23e)$$

(2.2.23b)作为关于$\frac{T_{1,\,x}}{v_x}$的二阶常微分方程,有形式解

$$\frac{T_{1,\,x}}{v_x} = -2 \cdot 4^{-n-s}\alpha(t)F^{-1}v_{xx} \qquad (2.2.24)$$

其中$F^{-1} = \cos v\partial^{-1}\cos v\partial^{-1} + \sin v\partial^{-1}\sin v\partial^{-1}$是 mKdV 递推算子$F = \partial^2 + 4\partial u\partial^{-1}u,\ u = \frac{i}{2}v_x$的逆算子, 则由(2.2.23c, d, e) 推出,

$$\frac{T_{j,\,x}}{v_x} = 4 \cdot F^{-1} \cdot \frac{T_{j-1,\,x}}{v_x},\ (j = 2, \cdots, s,\ s+2, \cdots, n+s)$$

$$(2.2.25a)$$

$$\frac{T_{s+1,\,x}}{v_x} = 4 \cdot F^{-1} \cdot \frac{T_{s,\,x}}{v_x} - 2(4^{-1-n}\beta(t))F^{-1}(xv_x)_x$$

$$(2.2.25b)$$

代入(2.2.23a) 算得 sine-Gordon 系统的方程族

$$v_{tx} = \alpha(t)F^{-(n+s)}v_{xx} + \beta(t)F^{-n}(xv_x)_x \qquad (2.2.26)$$

定理 2.3 设$\alpha(t)$, $\beta(t)$是关于时间t的任意函数,如果谱参数λ随时间的变化规律为

$$\lambda_t = \frac{\beta(t)}{2}(4\lambda)^{1-n}$$

待定函数 A, B 满足边值条件 $A\mid_{u=0}=0$, $B\mid_{u=0}=\alpha(t)(4\lambda)^{-n-s}+\beta(t)$ $(4\lambda)^{-n}x$, 则有 sine-Gordon 系统的方程族

$$v_{xt}=\alpha(t)F^{-(n+s)}v_{xx}+\beta(t)F^{-n}(xv_x)_x$$

特别地, (1) 当 $\alpha(t)=1$, $\beta(t)=0$, 有

$$v_{xt}=F^{-(n+s)}v_{xx} \qquad (2.2.27a)$$

即为等谱 sine-Gordon 方程族[27, 79, 80]; 第一个非平凡方程为

$$v_{xt}=\sin v \qquad (2.2.27b)$$

(2) 当 $\alpha(t)=0$, $\beta(t)=1$, 有

$$v_{xt}=F^{-n}(xv_x)_x \qquad (2.2.28a)$$

即为非等谱 sine-Gordon 方程族[27, 71, 80]; 第一个非平凡方程为

$$v_{xt}=(\cos\partial^{-1}\cos v\partial^{-1}+\sin v\partial^{-1}\sin v\partial^{-1})(xv_x)_x$$
$$(2.2.28b)$$

(3) 当 $\alpha(t)=(2s+1)t$, $\beta(t)=1$, 有

$$v_{xt}=(2s+1)tF^{-(n+s)}v_x+F^{-n}(xv_x)_x \qquad (2.2.29a)$$

即为 sine-Gordon 系统的 τ 方程族[77].

当 $s=0$, $n=1$, τ 方程为

$$v_{xt}=t\sin v+(\cos v\partial^{-1}\cos\partial^{-1}+\sin v\ \partial^{-1}\sin v\partial^{-1})(xv_x)_x$$
$$(2.2.29b)$$

§2.3 AKNS 系统的方程族的导出及约化

§2.3.1 AKNS 系统方程族的导出

考虑矩阵谱问题

$$\phi_x = M\phi, \ M = \begin{pmatrix} -ik & q \\ r & ik \end{pmatrix}, \ \phi = \begin{pmatrix} \phi_1 \\ \phi_2 \end{pmatrix} \tag{2.3.1a}$$

和时间发展式

$$\phi_t = N\phi, \ N = \begin{pmatrix} A & B \\ C & -A \end{pmatrix} \tag{2.3.1b}$$

其中 q, r 是一对光滑的位势，k 是谱参数且满足 $k_t = -\dfrac{i}{2}\beta(t)$ $(2ik)^n$；而 A, B 与 C 是变量 t, x，位势 q, r 和谱参数 k 的待定函数，满足边界条件

$$N\big|_{(q, r)=(0, 0)}$$

$$= \begin{pmatrix} -\dfrac{\alpha(t)}{2}(2ik)^{m+n-1} - \dfrac{\beta(t)}{2}(2ik)^n x & 0 \\ 0 & \dfrac{\alpha(t)}{2}(2ik)^{m+n-1} + \dfrac{\beta(t)}{2}(2ik)^n x \end{pmatrix}$$

$$\tag{2.3.1c}$$

由(2.3.1)所满足的零曲率方程

$$M_t - N_x = [N, M] \tag{2.3.2}$$

给出

$$-ik_t - A_x + qC - rB = 0 \tag{2.3.3a}$$

$$q_t - B_x - 2ikB - 2qA = 0 \tag{2.3.3b}$$

$$r_t - C_x + 2ikC + 2rA = 0 \tag{2.3.3c}$$

由(2.3.3a)结合 N 所满足的边值条件(2.3.1c)推知

$$A = \partial^{-1}(r, q)\begin{pmatrix} -B \\ C \end{pmatrix} - \frac{1}{2}\beta(t)(2ik)^n x - \frac{1}{2}\alpha(t)(2ik)^{m+n-1}$$

$$\tag{2.3.4}$$

将(2.3.4)代入到(2.3.3b, c),有

$$\binom{q}{r}_t = L\binom{-B}{C} - 2ik\binom{-B}{C} + \alpha(t)(2ik)^{m+n-1}\sigma\binom{q}{r} + \beta(t)(2ik)^n\sigma\binom{xq}{xr}$$

$$(2.3.5a)$$

式中

$$L = \sigma\partial + 2\binom{q}{-r}\partial^{-1}(r, q), \quad \sigma = \begin{pmatrix} -1 & 0 \\ 0 & 1 \end{pmatrix} \quad (2.3.5b)$$

假设 B, C 是 k 的 $m+n-2$ 次多项式

$$\binom{B}{C} = \sum_{j=1}^{m+n-1}\binom{b_j}{c_j}(2ik)^{m+n-1-j} \qquad (2.3.6)$$

将(2.3.6)代入(2.3.5a)并令 $2ik$ 的各次幂系数相等有

$$\binom{q}{r}_t = L\begin{bmatrix} -b_{m+n-1} \\ c_{m+n-1} \end{bmatrix} \qquad (2.3.7a)$$

$$\begin{bmatrix} -b_j \\ c_j \end{bmatrix} = L\begin{bmatrix} -b_{j-1} \\ c_{j-1} \end{bmatrix} \quad (j = 2, 3, \cdots, m+n-1, j \neq m)$$

$$(2.3.7b)$$

$$\begin{bmatrix} -b_m \\ c_m \end{bmatrix} = L\begin{bmatrix} -b_{m-1} \\ c_{m-1} \end{bmatrix} + \beta(t)\binom{-xq}{xr} \qquad (2.3.7c)$$

$$\binom{b_1}{c_1} = \alpha(t)\binom{-q}{r} \qquad (2.3.7d)$$

由(2.3.7b, c, d)递推得

$$\begin{bmatrix} -b_{m+n-1} \\ c_{m+n-1} \end{bmatrix} = \alpha(t)L^{m+n-2}\binom{-q}{r} + \beta(t)L^{n-1}\binom{-xq}{xr} \quad (2.3.8)$$

将(2.3.8)代入(2.3.7a)给出 AKNS 系统的方程族

$$\binom{q}{r}_t = \alpha(t)L^{m+n-1}\binom{-q}{r} + \beta(t)L^n\binom{-xq}{xr} \quad (m = 1, 2, \cdots; n = 0, 1, \cdots)$$

$$(2.3.9)$$

如果 $\beta(t) = 0$，这时谱参数不随时间变化，(2.3.9)化为等谱方程族

$$\binom{q}{r}_t = \alpha(t)L^{m+n-1}\binom{-q}{r} \qquad (2.3.10a)$$

如果 $\alpha(t) = 0$，则有非等谱方程族

$$\binom{q}{r}_t = \beta(t)L^n\binom{-xq}{xr} \qquad (2.3.10b)$$

因此，我们称 $K_j = L^{j-1}K_1$, $(j = 1, 2, \cdots)$, $K_1 = \binom{-q}{r}$ 为

AKNS 等谱流，$\sigma_j = L^{j-1}\sigma_1$, $(j = 1, 2, \cdots)$, $\sigma_1 = \binom{-xq}{xr}$ 为 AKNS

非等谱流.

定理 2.4 设 $\alpha(t)$ 与 $\beta(t)$ 是时间 t 的任意函数，如果 k 随 t 的变化规律为

$$k_t = -\frac{i}{2}\beta(t)(2ik)^n$$

而待定矩阵 N 满足边值条件(2.3.2)，则可以唯一地确定出

$$\binom{B}{C} = \alpha(t)\sum_{j=1}^{m+n-1}(2ik)^{m+n-1-j}\widetilde{L}^{j-1}\binom{q}{r} + \beta(t)\sum_{j=1}^{n}(2ik)^{n-j}\widetilde{L}^{j-1}\binom{xq}{xr}$$

$$(2.3.11)$$

且 AKNS 系统的方程族为

$$\binom{q}{r}_t = \alpha(t)L^{m+n-1}\binom{-q}{r} + \beta(t)L^n\binom{-xq}{xr}$$

式中 L 是 AKNS 递推算子, $\tilde{L} = \sigma L \sigma$.

特别地,(1) 当 $\alpha(t) = 1$, $\beta(t) = 0$ 时, 即为等谱 AKNS 方程族 [27, 79, 80]

$$\binom{q}{r}_t = L^{m+n-1} \binom{-q}{r} \qquad (2.3.12a)$$

其前两个非平凡的方程为

$$\binom{q}{r}_t = \begin{bmatrix} -q_{xx} + 2q^2 r \\ r_{xx} - 2qr^2 \end{bmatrix} \qquad (2.3.12b)$$

$$\binom{q}{r}_t = \begin{bmatrix} q_{xxx} - 6qrq_x \\ r_{xxx} - 6qrr_x \end{bmatrix} \qquad (2.3.12c)$$

(2) 当 $\alpha(t) = 0$, $\beta(t) = 1$ 时, 即为非等谱 AKNS 方程族[27, 71, 80]

$$\binom{q}{r}_t = L^n \binom{-xq}{xr} \qquad (2.3.13a)$$

其前两个非平凡的方程为

$$\binom{q}{r}_t = xK_2 + 2\sigma K_1 - 2\sigma \binom{q}{r} \partial^{-1} qr \qquad (2.3.13b)$$

$$\binom{q}{r}_t = xK_3 + 3\sigma K_2 - 2K_1 \partial^{-1} qr + 4 \begin{bmatrix} q\partial^{-1} qr_x \\ r\partial^{-1} rq_x \end{bmatrix}$$
$$(2.3.13c)$$

(3) 当 $\alpha(t) = mt$, $\beta(t) = 1$ 时, 即为 AKNS 系统的 τ 方程族[77]

$$\binom{q}{r}_t = mtL^{m+n-1} \binom{-q}{r} + L^n \binom{-xq}{xr} \qquad (2.3.14a)$$

当 $n = 0$, $m = 3$ 时, τ 对称方程为

$$\binom{q}{r}_t = 3t \begin{bmatrix} -q_{xx} + 2q^2 r \\ r_{xx} - 2qr^2 \end{bmatrix} + \begin{pmatrix} -xq \\ xr \end{pmatrix} \qquad (2.3.14b)$$

§2.3.2 AKNS 系统的方程族的约化

§2.3.2.1 约化为 mKdV 系统的方程族

当 $(q, r) = (v, \mp v)$ 时，取奇数阶 AKNS 系统的方程族

$$\binom{q}{r}_t = \alpha(t) L^{2n+2m+1} \binom{-q}{r} + \beta(t) L^{2n+1} \binom{-xq}{xr}, \ (m = 0, 1, \cdots; \ n = 1, 2, \cdots,)$$
$$(2.3.15)$$

依次可得

$$L \begin{pmatrix} -v \\ \mp v \end{pmatrix} = \begin{bmatrix} v_x \\ \mp v_x \end{bmatrix} \qquad (2.3.16a)$$

$$L^3 \begin{pmatrix} -v \\ \mp v \end{pmatrix} = L^2 \begin{bmatrix} v_x \\ \mp v_x \end{bmatrix} = \begin{bmatrix} F v_x \\ \mp F v_x \end{bmatrix} \qquad (2.3.16b)$$

$$\cdots$$

和

$$L \begin{pmatrix} -xv \\ \mp xv \end{pmatrix} = \begin{bmatrix} (xv)_x \\ \mp (xv)_x \end{bmatrix} \qquad (2.3.16c)$$

$$L^3 \begin{pmatrix} -xv \\ \mp xv \end{pmatrix} = L^2 \begin{bmatrix} (xv)_x \\ \mp (xv)_x \end{bmatrix} = \begin{bmatrix} F(xv)_x \\ \mp F(xv)_x \end{bmatrix} \qquad (2.3.16d)$$

$$\cdots$$

其中 $F = \partial^2 \pm 4v^2 \pm 4v_x \partial^{-1} v$ 是 mKdV 递推算子.

一般地

$$L^{2m+2n+1} \begin{pmatrix} -v \\ \mp v \end{pmatrix} = \begin{bmatrix} F^{m+n}v_x \\ \mp F^{m+n}v_x \end{bmatrix} \tag{2.3.17a}$$

和

$$L^{2n+1} \begin{pmatrix} -xv \\ \mp xv \end{pmatrix} = \begin{bmatrix} F^n(xv)_x \\ \mp F^n(xv)_x \end{bmatrix} \tag{2.3.17b}$$

所以(2.3.15)就约化为 mKdV 系统的方程族(2.2.11),即:

$$v_t = \alpha(t)F^{n+m}v_x + \beta(t)F^n(xv)_x$$

§2.3.2.2 约化为非线性 Schrödinger 系统的方程族

在谱问题(2.3.1a)中,取 $r = \mp q^*$,$k_t = \dfrac{\beta(t)}{2}(2k)^n$ 及 N 满足边值条件

$$N \big|_{(q, r)=(0, 0)}$$

$$= \begin{bmatrix} -\dfrac{i}{2}\alpha(t)(2k)^{m+n-1} - \dfrac{i}{2}\beta(t)(2k)^n x & 0 \\[3mm] 0 & \dfrac{i}{2}\alpha(t)(2k)^{m+n-1} + \dfrac{i}{2}\beta(t)(2k)^n x \end{bmatrix} \tag{2.3.18}$$

直接算得非线性 Schrödinger 系统的方程族

$$\begin{bmatrix} q \\ \mp q^* \end{bmatrix}_t = (-i)^{m+n-2}\alpha(t)L^{m+n-1} \begin{bmatrix} -q \\ \mp q^* \end{bmatrix} + (-i)^{n-1}\beta(t)L^n \begin{bmatrix} -xq \\ \mp xq^* \end{bmatrix}$$

$$(m = 1, 2, \cdots; n = 0, 1, \cdots) \tag{2.3.19}$$

和

$$\begin{pmatrix} B \\ C \end{pmatrix} = \alpha(t) \sum_{j=1}^{m+n-1} (2k)^{m+n-1-j}(-i\tilde{L})^{j-1} \begin{pmatrix} q \\ r \end{pmatrix} + \beta(t) \sum_{j=1}^{n} (2k)^{n-j}(-i\tilde{L})^{j-1} \begin{pmatrix} xq \\ xr \end{pmatrix}$$

$$\tag{2.3.20}$$

式中 $L = \sigma\partial + 2\begin{pmatrix} q \\ \pm q^* \end{pmatrix}\partial^{-1}(\mp q^*, q)$, $\tilde{L} = \sigma L \sigma$.

§2.3.2.3　约化为 sine-Gordon 系统的方程族

sine-Gordon 方程所对应的谱问题也是 $q = -r = \dfrac{u_x}{2}$ 时的谱问题 $(2.3.1a, b)$，以 $-(2n+1)$ 代替 n, $2(n-s)+1$ 代替 m. 这时 $(2.3.8)$ 的负奇数阶 AKNS 系统的方程族

$$\begin{pmatrix} q \\ r \end{pmatrix}_t = \alpha(t)L^{-(2s+1)}\begin{pmatrix} -q \\ r \end{pmatrix} + \beta(t)L^{-(2n+1)}\begin{pmatrix} -xq \\ xr \end{pmatrix}, \ (n, s = 0, 1, 2, \cdots)$$

$$(2.3.21)$$

可约化得 sine-Gordon 系统的方程族

$$u_{xt} = \alpha(t)F^{-s-1}u_{xx} + \beta(t)F^{-n-1}(xu_x)_x, \ (n, s = 0, 1, \cdots)$$

$$(2.3.22)$$

式中 $F^{-1} = \cos u\,\partial^{-1}\cos u\,\partial^{-1} + \sin u\,\partial^{-1}\sin u\,\partial^{-1}$ 是 mKdV 递推算子的逆算子.

当 $\alpha(t) = [2(n-s)+1]t, \beta(t) = 1$ 时，即得 sine-Gordon 系统的 τ 方程族

$$u_{xt} = (2(n-s)+1)tF^{-s-1}u_{xx} + F^{-n-1}(xu_x)_x \quad (2.3.23)$$

第三章　KdV 系统的方程族的解

考虑 KdV 系统的方程族(2.1.10)的初值问题

$$u_t = \alpha(t) T^{n+s-1} u_x + \beta(t) T^n (x u_x + 2u) \tag{3.1a}$$

$$u(x, 0) = f(x) \tag{3.1b}$$

的解.

由第二章知道与之相联系的 Lax pair[79]是

$$\phi_{xx} + u\phi = \lambda\phi \tag{3.2a}$$

$$\phi_t = A\phi + B\phi_x \tag{3.2b}$$

其中

$$\lambda_t = \frac{1}{2}\beta(t)(4\lambda)^{n+1} \tag{3.2c}$$

$$B = \beta(t)(4\lambda)^n x + \alpha(t)(4\lambda)^{n+s-1} + 2\alpha(t) \sum_{j=1}^{n+s-1} (4\lambda)^{n+s-1-j} \partial^{-1} T^{j-1} u_x +$$

$$2\beta(t) \sum_{j=s}^{n+s-1} (4\lambda)^{n+s-1-j} \partial^{-1} T^{j-s}(x u_x + 2u) \tag{3.2d}$$

我们以下假定

(1) $u(x, t)$ 及 $u(x, t)$ 所需要的关于 x 的各阶导数,当 x 趋于无穷时充分快地趋于零.

(2) $u(x, t) \in P_\mu$.

(3) $\lambda = -k^2$.

§3.1 正散射问题

给定位势函数 u，研究线性问题(3.2a)的特征值 λ 以及特征函数 ϕ 的性质称为正散射问题.

§3.1.1 特征函数的性质

根据二阶线性常微分方程的理论容易得出，线性问题(3.2a)有两个线性无关的解 $\phi_1(x, k)$，$\phi_2(x, k)$ 且具有渐进性质：

$$\phi_1(x, k) \sim \mathrm{e}^{\mathrm{i}kx}, \ x \to \infty \tag{3.1.1a}$$

$$\phi_2(x, k) \sim \mathrm{e}^{-\mathrm{i}kx}, \ x \to -\infty \tag{3.1.1b}$$

物理上称这两个线性无关解为 Jost 函数. 它们对所有的 x 值有界，而且关于 k 在上半平面解析，且连续到实轴.

Jost 函数关于 x, k 的导函数具有渐进式

$$\phi_{1,x}(x, k) \sim \mathrm{i}k\mathrm{e}^{\mathrm{i}kx}, \ \phi_{1,k}(x, k) \sim \mathrm{i}x\mathrm{e}^{\mathrm{i}kx}, \ x \to \infty \tag{3.1.2a}$$

$$\phi_{2,x}(x, k) \sim -\mathrm{i}k\mathrm{e}^{-\mathrm{i}kx}, \ \phi_{2,k} \sim -\mathrm{i}x\mathrm{e}^{-\mathrm{i}kx}, \ x \to -\infty \tag{3.1.2b}$$

事实上，求解线性问题(3.2a)等价于求解积分方程

$$\phi_1(x, k) = \mathrm{e}^{\mathrm{i}kx} + \int_x^\infty \frac{\sin k(x-y)}{k} u(y)\phi_1(x, k)\mathrm{d}y \tag{3.1.3a}$$

$$\phi_2(x, k) = \mathrm{e}^{-\mathrm{i}kx} - \int_{-\infty}^x \frac{\sin k(x-y)}{k} u(y)\phi_2(x, k)\mathrm{d}y \tag{3.1.3b}$$

若令

$$f_1(x, k) = e^{-ikx}\phi_1(x, k), \quad f_2(x, k) = e^{ikx}\phi_2(x, k)$$

$$(3.1.4)$$

方程(3.1.3)可写成

$$f_1(x, k) = 1 - \int_x^\infty \frac{e^{-2ik(x-y)} - 1}{2ik} u(y) f_1(y, k) \mathrm{d}y$$

$$(3.1.5a)$$

$$f_2(x, k) = 1 - \int_{-\infty}^x \frac{e^{2ik(x-y)} - 1}{2ik} u(y) f_2(y, k) \mathrm{d}y$$

$$(3.1.5b)$$

按逐次逼近法对方程(3.1.5a)构造函数列

$$f_1^{(0)}(x, k) = 1 \qquad (3.1.6a)$$

$$f_1^{(j)}(x, k) = -\int_x^\infty \frac{e^{-2ik(x-y)} - 1}{2ik} u(y) f_1^{(j-1)}(y, k) \mathrm{d}y \quad (j = 1, 2, \cdots)$$

$$(3.1.6b)$$

因为当 $\mathrm{Im}\, k \geqslant 0$ 时，积分中的指数函数始终保持有界，且有估计式

$$\frac{e^{-2ik(x-y)} - 1}{2ik} = \int_0^{y-x} e^{2iks} \mathrm{d}s \leqslant y - x \qquad (3.1.7)$$

所以通过归纳法可得

$$|f_1^{(j)}(x, k)| \leqslant \int_x^\infty (y-x) |u(y)| |f_1^{(j-1)}(y, k)| \mathrm{d}y \leqslant \frac{\sigma^j(x)}{j!}$$

$$(3.1.8a)$$

式中

$$\sigma(x) = \int_x^\infty (y-x) |u(y)| \mathrm{d}y \qquad (3.1.8b)$$

由此可见函数项级数

$$f_1^{(0)}(x, k) + f_1^{(1)}(x, k) + \cdots + f_1^{(j)}(x, k) + \cdots \quad (3.1.9)$$

当 $k \neq 0$ 时在半无界区间 $a \leqslant x < \infty$ 上绝对一致收敛,它所确定的
和函数 $f_1(x, k)$ 在 k 的上半平面直到实轴都是连续的且满足积分方
程(3.1.5a)和估计式

$$|f_1(x, k)| \leqslant e^{\sigma(x)} \quad (3.1.10)$$

若方程(3.1.5a)存在两个解 $f_1(x, k)$ 及 $\tilde{f}_1(x, k)$,则其差满足
不等式

$$|f_1(x, k) - \tilde{f}_1(x, k)| \leqslant \int_x^\infty (y-x)|u(y)||f_1(y, k) - \tilde{f}_1(y, k)| \, \mathrm{d}y$$
$$(3.1.11)$$

令

$$F(x, k) = |f_1(x, k) - \tilde{f}_1(x, k)| \quad (3.1.12a)$$

$$G(x, k) = \int_x^\infty (y-x)|u(y)|F(y, k)\mathrm{d}y \quad (3.1.12b)$$

(3.1.11)可简写成

$$F(x, k) \leqslant G(x, k) \quad (3.1.13)$$

将(3.1.12b)对 x 求导并利用(3.1.13)得

$$G_x(x, k) \geqslant -\int_x^\infty |u(y)|G(y, k)\mathrm{d}y \geqslant -\left(\int_x^\infty |u(y)| \, \mathrm{d}y\right)G(x, k)$$
$$= \left(\frac{\mathrm{d}}{\mathrm{d}x}\int_x^\infty (y-x)|u(y)| \, \mathrm{d}y\right)G(x, k)$$
$$(3.1.14)$$

由此推得

$$(e^{-\sigma(x)}G(x, k))_x \geqslant 0 \quad (3.1.15)$$

其中 $\sigma(x)$ 是积分(3.1.8b). 这个不等式说明 $e^{-\sigma(x)}G(x, k)$ 关于 x 是
增函数,由于 $x \to \infty$ 时,此函数的极限为零,所以 $G(x, k)$ 恒等于零.

从而

$$f_1(x, k) = \tilde{f}_1(x, k) \qquad (3.1.16)$$

再回到积分方程(3.1.3a).我们就能得出,当 $k \neq 0$, $\text{Im} k \geqslant 0$ 时在半无界区间 $a \leqslant x < \infty$ 上,此方程存在唯一的解 $\phi_1(x, k)$,其在 k 的上半平面直到实轴是连续的.

应用类似的推理于积分方程(3.1.3b)可知解 $\phi_2(x, k)$ 当 $k \neq 0$, $\text{Im} k \geqslant 0$ 时在半无界 区间 $-\infty < x \leqslant b$ 上存在唯一,且在 k 的上半平面直到实轴均是连续的.

若 $k = 0$,这时与(3.2a)及(3.1.1)等价的积分方程是

$$\phi_1(x) = 1 + \int_x^\infty (x - y) u(y) \phi_1(y) \mathrm{d}y \qquad (3.1.17a)$$

$$\phi_2(x) = 1 - \int_{-\infty}^x (x - y) u(y) \phi_2(y) \mathrm{d}y \qquad (3.1.17b)$$

这里为简化起见分别以 $\phi_1(x)$ 及 $\phi_2(x)$ 记 $\phi_1(x, 0)$ 及 $\phi_2(x, 0)$.容易看出方程(3.1.17)恰是(3.1.3)当 k 趋于零的极限形式,从而不难得出对任意的 $a \leqslant x < \infty$(或 $-\infty < x \leqslant b$)这些方程解的存在唯一性.

类似的推理可以得到 Jost 函数关于 x 及 k 的可微性.

定义 3.1 $\phi_1(x, k)$ 和 $\phi_2(x, k)$ 构成的 wronski 行列式为

$$W(\phi_1(x, k), \phi_2(x, k)) = \phi_1(x, k)\phi_{2,x}(x, k) - \phi_{1,x}(x, k)\phi_2(x, k) \qquad (3.1.18)$$

则容易推知

$$W(\phi_1(x, k), \phi_1(x, -k)) = -2\mathrm{i}k \qquad (3.1.19)$$

事实上

$$\frac{\mathrm{d}}{\mathrm{d}x} W(\phi_1(x, k), \phi_1(x, -k)) = 0 \qquad (3.1.20)$$

因此 $W(\phi_1(x, k), \phi_1(x, -k))$ 是一个与 x 无关的常数,利用 Jost 函

数的渐进性质即得(3.1.19).

§3.1.2　反射系数与穿透系数

当 k 是实数时,Jost 函数组$\{\phi_1(x, k), \phi_1(x, -k)\}$和$\{\phi_2(x, k),$
$\phi_2(x, -k)\}$分别构成线性问题(3.2a)的基本解组,它们之间存在线性关系

$$\phi_2(x, k) = a(k)\phi_1(x, -k) + b(k)\phi_1(x, k) \quad (3.1.21a)$$

$$\phi_2(x, -k) = b(-k)\phi_1(x, -k) + a(-k)\phi_1(x, k) \quad (3.1.21b)$$

由(3.1.19),(3.1.21a)算得

$$a(k) = \frac{1}{2ik}W(\phi_2(x, k), \phi_1(x, k)) \quad (3.1.22a)$$

$$b(k) = \frac{1}{2ik}W(\phi_1(x, -k), \phi_2(x, k)) \quad (3.1.22b)$$

因此有

性质 1　$a(k)$在 k 的上半平面解析,且连续到实轴,而 $b(k)$仅在实轴 $\text{Im}k = 0$ 上有定义.

用 $a(k)$除(3.1.21a)两端,得

$$\frac{1}{a(k)}\phi_2(x, k) = \phi_1(x, -k) + \frac{b(k)}{a(k)}\phi_1(x, k) \quad (3.1.23)$$

物理上称 $T(k) = \frac{1}{a(k)}$ 为穿透系数,$R(k) = \frac{b(k)}{a(k)}$ 为反射系数.

如果 $\phi(x, k)$是(3.2a)的解,则 $\phi(x, -k)$也是(3.2a)的解,由解的唯一性,推知

$$\phi_1^*(x, k) = \phi_1(x, -k) = \phi_2(x, k) \quad (3.1.24a)$$

$$\phi_2^*(x, k) = \phi_2(x, -k) = \phi_1(x, k) \quad (3.1.24b)$$

所以

$$a^*(k) = a(-k), \quad b^*(k) = b(-k) \qquad (3.1.25a,\ b)$$

由(3.1.21)算得

$$|a(k)|^2 - |b(k)|^2 = 1 \qquad (3.1.25c)$$

§3.1.3 谱的分布

性质 2 Schrödinger 算子 $\partial^2 + u$ 是自共轭算子. 即 $(\partial^2 + u)^* = \partial^2 + u$.

事实上,对于任意两个可微函数 ϕ, ψ, 有

$$\langle (\partial^2 + u)\phi, \psi \rangle = \langle \phi, (\partial^2 + u)\psi \rangle = \langle \phi, (\partial^2 + u)^* \psi \rangle$$

定理 3.1 $a(k)$ 的零点 k_j 为简单零点,位于上半虚轴,即 $k_j = i\kappa_j$, $(\kappa_j > 0)$,且零点的个数为有限个.

证明 注意到(3.1.25c)在 k 是实数时成立,因此 $a(k)$ 在实轴没有零点,其零点只能分布在上半复平面,又根据性质 2 容易推知 $a(k)$ 的零点在虚轴上. 为了证明 $a(k)$ 的零点的个数为有限个,将 (3.1.21a)改写为

$$\phi_2(x, k) - a(k)\phi_1(x, -k) = b(k)\phi_1(x, k) \qquad (3.1.26)$$

令

$$\Delta = \phi_2(x, k) - a(k)\phi_1(x, -k) = b(k)\phi_1(x, k)$$

$$(3.1.27)$$

将(3.1.3)中 ϕ_1, ϕ_2 代入(3.1.27),得

$$\Delta = e^{-ikx} - a(k)e^{-ikx} - \int_{-\infty}^{\infty} \frac{\sin k(x-y)}{k} u(y)\phi_2(y, k)\mathrm{d}y +$$

$$\int_x^{\infty} \frac{\sin k(x-y)}{k} u(y)\Delta(y, k)\mathrm{d}y$$

$$\Delta = b(k)\mathrm{e}^{\mathrm{i}kx} + \int_x^\infty \frac{\sin k(x-y)}{k}u(y)b(k)\phi_1(y,\,k) = b(k)\mathrm{e}^{\mathrm{i}kx} +$$

$$\int_x^\infty \frac{\sin k(x-y)}{k}u(y)\Delta(y,\,k)\mathrm{d}y$$

上两式分别乘以 $\mathrm{e}^{\mathrm{i}kx}$，然后比较 1 与 $\mathrm{e}^{2\mathrm{i}kx}$ 的系数，得

$$a(k) = 1 + \frac{1}{2\mathrm{i}k}\int_{-\infty}^\infty u(y)\phi_2(y,\,k)\mathrm{e}^{\mathrm{i}ky}\mathrm{d}y \tag{3.1.28a}$$

$$b(k) = \frac{1}{2\mathrm{i}k}\int_{-\infty}^\infty u(y)\phi_2(y,\,k)\mathrm{e}^{\mathrm{i}ky}\mathrm{d}y \tag{3.1.28b}$$

由此可知

$$\text{当 } k \to \infty \text{ 时, } a(k) \sim 1 \tag{3.1.28c}$$

根据解析函数的性质，其零点个数必为有限个.

最后给出 $a(k)$ 的零点为简单零点，若 k_j 是 $a(k)$ 在上半平面的零点，则

$$W(\phi_2(x,\,k_j),\,\phi_1(x,\,k_j)) = 0$$

即 Jost 函数 $\phi_1(x,\,k_j)$，$\phi_2(x,\,k_j)$ 是线性相关的，存在常数 b_j，使得

$$\phi_2(x,\,k_j) = b_j\phi_1(x,\,k_j) \tag{3.1.29}$$

将方程

$$\phi_{1,xx}(x,\,k) + u\phi_1(x,\,k) = -k^2\phi_1(x,\,k) \tag{3.1.30}$$

对 k 求导，有

$$(\partial^2 + u + k^2)\phi_{1,k}(x,\,k) = -2k\phi_1(x,\,k) \tag{3.1.31}$$

以 $\phi_2(x,\,k)$ 乘(3.1.31)，得

$$[(\partial^2 + u + k^2)\phi_{1,k}(x,\,k)]\phi_2(x,\,k) = -2k\phi_1(x,\,k)\phi_2(x,\,k) \tag{3.1.32}$$

再以 $\phi_2(x,k)$ 代替(3.1.30)式的 $\phi_1(x,k)$ 后,与 $\phi_{1,k}(x,k)$ 相乘,给出

$$[(\partial^2+u+k^2)\phi_2(x,k)]\phi_{1,k}(x,k)=0 \qquad (3.1.33)$$

(3.1.32)减去(3.1.33),得

$$(W(\phi_2(x,k),\phi_{1,k}(x,k)))_x=-2k\phi_1(x,k)\phi_2(x,k) \qquad (3.1.34)$$

从 x 到 l 积分有

$$W(\phi_2(x,k),\phi_{1,k}(x,k))\Big|_x^l=2k\int_x^l\phi_1(y,k)\phi_2(y,k)\mathrm{d}y \qquad (3.1.35a)$$

类似计算可得

$$W(\phi_{2,k}(x,k),\phi_1(x,k))\Big|_{-l}^x=-2k\int_{-l}^x\phi_1(y,k)\phi_2(y,k)\mathrm{d}y \qquad (3.1.35b)$$

(3.1.35)两式相加并取 $k=\mathrm{i}\kappa_j$,令 $l\to\infty$,最后达到

$$a_k(\mathrm{i}\kappa_j)=-\mathrm{i}b_j\int_{-\infty}^{\infty}\phi_1^2(x,\mathrm{i}\kappa_j)\mathrm{d}x \qquad (3.1.36)$$

由于 $u(x)$ 是实函数,$\mathrm{i}\kappa_j$ 是纯虚数,因此由(3.2a)知 $\phi_1(x,\mathrm{i}\kappa_j)$ 也是实函数,所以 $a_k(\mathrm{i}\kappa_j)\neq0$,由此说明 $\mathrm{i}\kappa_j$ 是 $a(k)$ 的简单零点.

根据 Jost 函数的渐进性质和(3.1.29)知,若 k_j 为 $a(k)$ 的零点,则函数 $\phi_1(x,k_j)$ 无论 x 趋于正无穷还是趋于负无穷均以指数衰减而在整个实轴上成为平方可积.因此 $\phi_1(x,k_j)$ 是满足线性问题(3.2a)的特征函数,$-k_j^2$ 是相应的特征值,这样的特征值称为**离散谱**.此外,当 k 为实数时,$\phi(x,k)$ 在整个实轴上有界,而成为特征函数,这样的特征值称为连续谱.

定义 3.2 设 k_j 是给定的线性问题(3.2a)的特征值,$\phi(x,k_j)$ 是

相应的特征函数,若存在常数 c_j 使得

$$\int_{-\infty}^{\infty} c_j^2 \phi^2(x, k_j)\mathrm{d}x = 1 \qquad (3.1.37)$$

则 c_j 称为本征函数 $\phi(x, k_j)$ 的归一化因子,而称 $c_j\phi(x, k_j)$ 为归一化本征函数

由(3.1.36)易见

$$c_j^2 = \frac{-\mathrm{i}b_j}{a_k(\mathrm{i}\kappa_j)} \qquad (3.1.38)$$

由上可见线性问题(3.2a)具有有限个特征函数为平方可积的离散谱 $\mathrm{i}\kappa_j$ 和特征函数为非平方可积的充满实轴的连续谱 k.

定义 3.3 称 $S(t)$:

$$\left\{R(k) = \frac{a(k)}{b(k)}, \ \mathrm{Im}\,k = 0, \ \{\mathrm{i}\kappa_j, c_j\}, \ \kappa_j > 0, \ (j = 1, 2, \cdots, l)\right\}$$

$$(3.1.39)$$

为线性问题(3.2a)的散射数据.

§3.2 反散射问题

已知线性问题(3.2a)的散射数据,恢复位势 $u(x)$,即为反散射问题.

§3.2.1 平移变换与 GLM 积分方程

GLM(Gel'fand-Levi tan-Marchenkò)积分方程和散射数据随时间的演化关系,是散射反演的关键. 这里简要介绍 GLM 积分方程的由来.

一、平移变换
求给定位势 $u(x)$ 的线性问题(3.2a)的解,等价于求解积分方程

组(3.1.3). 如果取 $u(x)=0$, 则 $\phi_1(x,k)=\mathrm{e}^{ikx}$, $\phi_2(x,k)=\mathrm{e}^{-ikx}$ 是线性问题(3.2a)的两个基本解. 则存在一个平移变换, 把 $u(x)=0$ 时线性问题(3.2a)的解 e^{ikx}, e^{-ikx} 变为 $u(x)\neq 0$ 时线性问题(3.2a)的解 $\phi_1(x,k)$, $\phi_2(x,k)$, 即:

$$\phi_1(x,k)=\mathrm{e}^{ikx}+\int_x^\infty K_1(x,y)\mathrm{e}^{iky}\mathrm{d}y \qquad (3.2.1a)$$

$$\phi_2(x,k)=\mathrm{e}^{-ikx}+\int_{-\infty}^x K_2(x,y)\mathrm{e}^{-iky}\mathrm{d}y \qquad (3.2.1b)$$

事实上将(3.2.1)代入(3.1.3a)给出

$$K_1(x,x+y)=-\frac{1}{2}\int_{x+\frac{y}{2}}^\infty u(s)\mathrm{d}s-\frac{1}{2}\int_0^y\int_{x+\frac{y-z}{2}}^\infty u(s)K_1(s,s+z)\mathrm{d}s\mathrm{d}z$$

$$(3.2.2)$$

此即为核 $K_1(x,x+y)$ 所满足的积分方程. 当 y 为零时

$$K_1(x,x)=-\frac{1}{2}\int_x^\infty u(s)\mathrm{d}s \qquad (3.2.3)$$

依照逐次逼近法作函数列

$$K_1^{(0)}(x,x+y)=-\frac{1}{2}\int_{x+\frac{y}{2}}^\infty u(s)\mathrm{d}s \qquad (3.2.4a)$$

$$K_1^{(j)}(x,x+y)=-\frac{1}{2}\int_0^y\int_{x+\frac{y-z}{2}}^\infty u(s)K_1^{(j-1)}(s,s+z)\mathrm{d}s\mathrm{d}z$$

$$(3.2.4b)$$

其中 $y>0$, $j\geqslant 1$. 容易对这些函数作出估计

$$|K_1^{(0)}(x,x+y)|\leqslant\frac{1}{2}\int_x^\infty|u(s)|\mathrm{d}s \qquad (3.2.5a)$$

$$|K_1^{(j)}(x,x+y)|\leqslant\frac{1}{2\cdot j!}\sigma^j(x)\int_x^\infty|u(s)|\mathrm{d}s \qquad (3.2.5b)$$

式中 $\sigma(x)$ 表示为 (3.1.8b). 由此推知无穷级数

$$K_1^{(0)}(x,\ x+y) + K_1^{(1)}(x,\ x+y) + \cdots + K_1^{(j)}(x,\ x+y) + \cdots \tag{3.2.6}$$

在区域 $a \leqslant x < \infty,\ y \geqslant 0$ 上绝对一致收敛,它的和即为方程 (3.2.2) 的解 $K_1(x,\ x+y)$,且当 $y = 0$ 时满足边值条件 (3.2.3).

如果方程 (3.2.2) 还存在另一个满足条件 (3.2.3) 的解 $\widetilde{K}_1(x,\ x+y)$,则对此二解的差应成立

$$K_1(x,\ x+y) - \widetilde{K}_1(x,\ x+y) = -\frac{1}{2}\int_0^y\int_{x+\frac{y-z}{2}}^\infty u(s)(K_1(s,\ s+z) -$$
$$\widetilde{K}_1(s,\ s+z))dsdz \tag{3.2.7}$$

令

$$F(x,\ y) = |\,K_1(x,\ x+y) - \widetilde{K}_1(x,\ x+y)\,| \tag{3.2.8a}$$

$$G(x,\ y) = \frac{1}{2}\int_0^y\int_x^\infty |\,u(s)\,| F(s,\ z)dsdz \tag{3.2.8b}$$

我们有

$$F(x,\ y) \leqslant G(x,\ y) \tag{3.2.9}$$

将 $G(x,\ y)$ 对 y 微商并利用此不等式得

$$G_y(x,\ y) \leqslant \frac{1}{2}\Big(\int_x^\infty |\,u(s)\,| ds\Big)G(x,\ y) \tag{3.2.10}$$

由此推知

$$(e^{-\frac{1}{2}y\int_x^\infty |u(s)|ds}G(x,\ y))_y \leqslant 0 \tag{3.2.11}$$

它说明括号内的非负函数关于 y 是递减的. 但因 y 为零时, $G(x,\ y)$ 为零,故对于大于零的 y, $G(x,\ y)$ 必恒等于零. 所以

$$K_1(x,\ x+y) = \widetilde{K}(x,\ x+y) \tag{3.2.12}$$

类似的分析可知 $K_2(x, y)$ 也是存在唯一的. 现在进而说明 $K_1(x, y)$ 及 $K_2(x, y)$ 对 x 及 y 均可微, 仍以 $K_1(x, y)$ 为例.

近似列(3.2.4)中的每一函数对 x 及 y 是可微的, 其导数为

$$K_{1,x}^{(0)}(x, x+y) = \frac{1}{2}u\left(x+\frac{y}{2}\right) \tag{3.2.13a}$$

$$K_{1,y}^{(0)}(x, x+y) = \frac{1}{4}u\left(x+\frac{y}{2}\right) \tag{3.2.13b}$$

$$K_{1,x}^{(j)}(x, x+y) = \frac{1}{2}\int_0^y u\left(x+\frac{y-z}{2}\right)K_1^{(j-1)}\left(x+\frac{y-z}{2}, x+\frac{y+z}{2}\right)dz \tag{3.2.14a}$$

$$K_{1,y}^{(j)}(x, x+y) = -\frac{1}{2}\int_x^\infty u(s)K_1^{(j-1)}(s, s+y)ds +$$

$$\frac{1}{4}\int_0^y u\left(x+\frac{y-z}{2}\right)K_1^{(j-1)}\left(x+\frac{y-z}{2}, x+\frac{y+z}{2}\right)dz \tag{3.2.14b}$$

借助 $K_1^{(j)}(x, x+y)$ 的估计式当 $j \geq 1$ 时不难算得

$$|K_{1,x}^{(j)}(x, x+y)| \leqslant \frac{1}{4}\frac{1}{(j-1)!}\sigma^{j-1}(x)\left(\int_x^\infty |u(s)|\,ds\right)^2 \tag{3.2.15a}$$

$$|K_{1,y}^{(j)}(x, x+y)| \leqslant \frac{1}{4}\frac{1}{(j-1)!}\sigma^{j-1}(x)\left(\int_x^\infty |u(s)|\,ds\right)^2 \tag{3.2.15b}$$

于是(3.2.6)逐项求导后所得的级数

$$K_{1,x}^{(0)}(x, x+y) + K_{1,x}^{(1)}(x, x+y) + \cdots + K_{1,x}^{(j)}(x, x+y) + \cdots \tag{3.2.16a}$$

$$K_{1,y}^{(0)}(x, x+y) + K_{1,y}^{(1)}(x, x+y) + \cdots + K_{1,y}^{(j)}(x, x+y) + \cdots$$

$$(3.2.16b)$$

在域 $a \leqslant x < \infty$, $y \geqslant 0$ 上绝对一致收敛，其和为核 $K_1(x, y)$ 对 x 与 y 的导数 $K_{1,x}(x, x+y)$ 与 $K_{1,y}(x, x+y)$. 特别当 $y = 0$ 时有

$$\frac{\mathrm{d}}{\mathrm{d}x} K(x, x) = \frac{1}{2} u(x) \qquad (3.2.17)$$

此乃说明已知 $K_1(x, x)$，则位势 $u(x)$ 可得到恢复.

二、GLM 积分方程

定理 3.2 设给定线性问题 (3.2a) 的散射数据

$$\{\mathrm{Im}\, k = 0, R(k), \mathrm{i}\kappa_j, c_j^2, (j = 1, 2, \cdots, l)\}$$

并引入函数

$$F_c(x) = \frac{1}{2\pi} \int_{-\infty}^{\infty} R(k) \mathrm{e}^{\mathrm{i}kx} \mathrm{d}k, \; F_d(x) = \sum_{j=1}^{l} c_j^2 \mathrm{e}^{-\kappa_j x}$$

$$(3.2.18a)$$

$$F(x) = F_c(x) + F_d(x) \qquad (3.2.18b)$$

则平移变换的积分核 $K(x, y) = K_1(x, y)$ 满足 GLM 方程

$$K(x, y) + F(x+y) + \int_x^{\infty} K(x, z) F(z+y) \mathrm{d}z = 0$$

$$(3.2.19)$$

证明 将 (3.1.23) 两端同时减去 $\mathrm{e}^{-\mathrm{i}kx}$，然后对 k 作 Fourier 变换，得

$$\int_{-\infty}^{\infty} (T(k)\phi_2(x, k) - \mathrm{e}^{-\mathrm{i}kx}) \mathrm{e}^{\mathrm{i}ky} \mathrm{d}k = \int_{-\infty}^{\infty} (\phi_1(x, -k) - \mathrm{e}^{-\mathrm{i}kx} +$$

$$R(k)\phi_1(x, k)) \mathrm{e}^{\mathrm{i}ky} \mathrm{d}k \qquad (3.2.20)$$

考察 (3.2.20) 式左端的积分. 因 $T(k)\phi_2(x, k) - \mathrm{e}^{-\mathrm{i}kx}$ 在上半平面有有限个简单极点且当 $k \to \infty$ 时趋于零，由残数定理推知

$$\int_{-\infty}^{\infty} (T(k)\phi_2(x,k) - e^{-ikx})e^{iky}\,dk = 2i\pi \sum_{j=1}^{l} \frac{\phi_2(x,i\kappa_j)}{a_k(i\kappa_j)} e^{-\kappa_j y}$$

$$(3.2.21)$$

利用(3.1.36)与(3.2.1a)算得

$$\frac{\phi_2(x,i\kappa_j)}{a_k(i\kappa_j)} = \frac{b_j\phi_1(x,i\kappa_j)}{a_k(i\kappa_j)} = ic_j^2\phi_1(x,i\kappa_j)$$

$$= ic_j^2\left(e^{-\kappa_j x} + \int_x^{\infty} K(x,z)e^{-\kappa_j z}\,dz\right) \quad (3.2.22)$$

将其代入(3.2.21)得(3.2.20)左端积分值为

$$-2\pi \sum_{j=1}^{l} c_j^2 e^{-\kappa_j(x+y)} - 2\pi \int_x^{\infty} K(x,z) \sum_{j=1}^{l} c_j^2 e^{-\kappa_j(y+z)}\,dz$$

$$(3.2.23)$$

其次,将(3.2.1a)代入(3.2.20)的右端给出

$$2\pi K(x,y) + \int_{-\infty}^{\infty} R(k)e^{ik(x+y)}\,dk + \int_x^{\infty} K(x,z)\left(\int_{-\infty}^{\infty} R(k)e^{ik(z+y)}\,dk\right)dz$$

$$(3.2.24a)$$

注意函数 $F_c(x)$ 及 $F_d(x)$ 的定义 (3.2.18a),积分等式 (3.2.20)可写成

$$-F_d(x+y) - \int_x^{\infty} K(x,z)F_d(z+y)\,dz = K(x,y) + F_c(x+y) +$$

$$\int_x^{\infty} K(x,z)F_c(z+y)\,dz$$

$$(3.2.24b)$$

此即为所要的方程(3.2.19).

§3.2.2 散射数据随时间的演化关系

这一节,我们导出散射数据随时间的演化规律. 它包括等谱 KdV

2005 年上海大学
博士学位论文 ■

方程族[15]、非等谱 KdV 方程族[20]两种情形.

引理 3.1 假设 $\phi(x, k)$ 是线性问题(3.2a)的一个解，A 和 B 满足相容性条件 $\phi_{xxt} = \phi_{txx}$，即：

$$2A_x + B_{xx} = 0, \ u_t = -A_{xx} - 2(\lambda - u)B_x + u_x B + \lambda_t$$

则有

$$P(x, k) = \phi_t(x, k) - A\phi(x, k) - B\phi_x(x, k) \quad (3.2.25)$$

也是线性问题(3.2a)的解.

引理 3.2 $(\phi^2)_x$ 是算子 T 关于特征值 4λ 的特征函数.

引理 3.3 $\partial^{-1}T = \partial + 2\partial^{-1}u + 2u\partial^{-1}$ 是一个反对称算子.

定理 3.3 线性问题(3.2a)的散射数据满足下列演化规律

$$\kappa_j(t) = \frac{1}{\left(\kappa_j^{-2n}(0) - 2n \cdot 4^n \int_0^t \beta(z)\mathrm{d}z\right)^{\frac{1}{2n}}} \quad (3.2.26a)$$

$$c_j(t) = c_j(0)\mathrm{e}^{\int_0^t \left[4^n\left(n+\frac{1}{2}\right)\beta(z)\kappa_j^{2n}(z) - 4^{n+s-1}\alpha(z)\kappa_j^{2n+2s-1}(z)\right]\mathrm{d}z}$$

$$(3.2.26b)$$

$$a(t, k(t)) = a(0, k(0)), \ b(t, k(t)) = b(0, k(0))\mathrm{e}^{\int_0^t \mathrm{i}k(s)a(s)(-4k^2(s))^{n+s-1}\mathrm{d}s}$$

$$(3.2.26c)$$

其中 $j = 1, 2, \cdots, l$，$\{\mathrm{i}\kappa_j(0), c_j(0), a(0, k(0)), b(0, k(0))\}$ 是 (3.2a) 当 $u(t, x) = u(0, x)$ 的散射数据.

特别地，当 $n = 0$ 时，对(3.2.26)取极限即可得到相应的演化规律.

证明 设 $\phi(x, k)$ 是线性问题(3.2a)的一个非零解，$\bar{\phi}(x, k)$ 是与 $\phi(x, k)$ 线性无关的解，则存在常数 γ_1，γ_2 使得下式成立

$$\phi_t(x, k) - A\phi(x, k) - B\phi_x(x, k) = \gamma_1\phi(x, k) + \gamma_2\bar{\phi}(x, k)$$

$$(3.2.27)$$

首先设 k 为离散谱 $\mathrm{i}\kappa_j(\kappa_j > 0)$，在(3.2.27)中选定 $\phi(x, \mathrm{i}\kappa_j)$ 为

对应的归一化特征函数，即：$\phi(x, i\kappa_j) = c_j\phi_1(x, i\kappa_j)$，$c_j$ 是 $\phi_1(x, i\kappa_j)$ 的归一化常数.

令 x 趋于正无穷时，由于 $\phi(x, i\kappa_j)$ 以指数衰减，则 $\tilde{\phi}(x, i\kappa_j)$ 必以指数增长，知 $\gamma_2 = 0$，于是(3.2.27)化为

$$\phi_t(x, i\kappa_j) - A\phi(x, i\kappa_j) - B\phi_x(x, i\kappa_j) = \gamma_1\phi(x, i\kappa_j)$$

$$(3.2.28)$$

为了求出 γ_1，以 $2\phi(x, i\kappa_j)$ 乘(3.2.28)式，注意到 $A = -\dfrac{1}{2}B_x$，所以

$$2\phi(x, i\kappa_j)\phi_t(x, i\kappa_j) + B_x\phi^2(x, i\kappa_j) - 2B\phi(x, i\kappa_j)\phi_x(x, i\kappa_j)$$

$$= 2\gamma_1\phi^2(x, i\kappa_j) \qquad (3.2.29)$$

进而有

$$\frac{\mathrm{d}}{\mathrm{d}t}\int_{-\infty}^{\infty}\phi^2(x, i\kappa_j)\mathrm{d}x + \int_{-\infty}^{\infty}[B_x\phi^2(x, i\kappa_j) - B(\phi^2(x, i\kappa_j))_x]\mathrm{d}x$$

$$= 2\gamma_1\int_{-\infty}^{\infty}\phi^2(x, i\kappa_j)\mathrm{d}x \qquad (3.2.30)$$

由于 $\displaystyle\int_{-\infty}^{\infty}\phi^2(x, i\kappa_j)\mathrm{d}x = 1$ 且 $x \to \pm\infty$ 时，$B\phi^2(x, i\kappa_j) \to 0$，可得

$$\gamma_1 = -\int_{-\infty}^{\infty}B(\phi^2(x, i\kappa_j))_x\mathrm{d}x \qquad (3.2.31)$$

这里重新定义 $\partial^{-1} = \dfrac{1}{2}\left(\displaystyle\int_{-\infty}^{x}\mathrm{d}x - \int_{x}^{\infty}\mathrm{d}x\right)$，并代入 B 的表达式(3.3d)给出

$$\gamma_1 = -<B, (\phi^2)_x>$$

$$= -<\alpha(t)(4\kappa_j^2)^{n+s-1}, (\phi^2)_x> - \sum_{j=1}^{n+s-1}2\alpha(t) \cdot (4\kappa_j^2)^{n+s-1-j}$$

$$= <\partial^{-1}T^{j-1}u_x, (\phi^2)_x> - <(4\kappa_j^2)^n\beta(t)x, (\phi^2)_x>$$

$$= \sum_{j=s}^{n+s-1} 2\beta(t)(4\kappa_j^2)^{n+s-1-j} <\partial^{-1}T^{j-s}(xu_x+2u), (\phi^2)_x>$$

$$(3.2.32)$$

由引理 3.2 与引理 3.3，有

$$<\partial^{-1}T^{j-s}(xu_x+2u), (\phi^2)_x> = <\partial^{-1}(xu_x+2u), T^{j-s}(\phi^2)_x>$$

$$= (2\kappa_j)^{2(j-s)} <\partial^{-1}(xu_x+2u), (\phi^2)_x> = -\frac{1}{2}(2\kappa_j)^{2(j-s+1)}$$

将其代入(3.2.32)，算得

$$\gamma_1 = (n+1)\beta(t)(2\kappa_j)^{2n}$$

从而(3.2.28) 可写成

$$\phi_t(x, i\kappa_j) + \frac{1}{2}B_x\phi(x, i\kappa_j) - B\phi_x(x, i\kappa_j)$$

$$= (n+1)\beta(t)(2\kappa_j)^{2n}\phi(x, i\kappa_j) \qquad (3.2.33)$$

在(3.2.33)式中令 $x \to +\infty$，即得

$$c_{j,t} - x\kappa_{j,t}c_j + \frac{1}{2}\beta(t)(2\kappa_j)^{2n}c_j + x\beta(t)\kappa_j(2\kappa_j)^{2n}c_j + \kappa_j\alpha(t)(2\kappa_j)^{2n+2s-2}$$

$$= (n+1)\beta(t)(2\kappa_j)^{2n}c_j \qquad (3.2.34)$$

由此推知

$$\kappa_{j,t}(t) = 4^n\beta(t)\kappa_j^{2n+1}(t),$$

$$c_{j,t}(t) = \left(4^n\left(n+\frac{1}{2}\right)\beta(t)\kappa_j^{2n}(t) - 4^{n+s-1}\alpha(t)\kappa_j^{2n+2s-1}(t)\right)c_j(t).$$

解此方程得到(3.2.26a, b).

其次设 k 是连续谱 $\text{Im}k = 0$，在(3.2.27)中选定 $\phi(x, k)$ 为 Jost 函数 $\phi_2(x, k)$，则(3.2.27)可以写成

$$\phi_t(x, k) - A\phi(x, k) - B\phi_x(x, k) = \gamma_1 \phi_2(x, -k) + \gamma_2 \phi(x, k)$$
$$(3.2.35)$$

令 x 趋于无穷求得

$$-\mathrm{i} k_t x\, \mathrm{e}^{-\mathrm{i}kx} + \frac{1}{2}\beta(t)(-4k^2)^n \mathrm{e}^{-\mathrm{i}kx} + \mathrm{i}k[\beta(t)(-4k^2)^n x +$$

$$\alpha(t)(-4k^2)^{n+s-1}]\mathrm{e}^{-\mathrm{i}kx} = \gamma_1 \mathrm{e}^{\mathrm{i}kx} + \gamma_2 \mathrm{e}^{-\mathrm{i}kx} \qquad (3.2.36)$$

比较各项系数给出

$$k_t = (-1)^n 4^n \beta(t) k^{2n+1}, \ \gamma_1 = 0, \ \gamma_2 = (-1)^n \frac{1}{2}\beta(t)(2k)^{2n} +$$

$$(-1)^{n+s-1}\alpha(t)(2k)^{2(n+s-1)}$$

如果用(3.1.21a)的右端来代替(3.2.35)中的 $\phi(x, k)$，并令 x 趋于正无穷大又得

$$\frac{\mathrm{d}a(t, k)}{\mathrm{d}t}\mathrm{e}^{-\mathrm{i}kx} - \mathrm{i}k_t xa(t, k)\mathrm{e}^{-\mathrm{i}kx} + \frac{\mathrm{d}b(t, k)}{\mathrm{d}t}\mathrm{e}^{\mathrm{i}kx} + \mathrm{i}k_t xb(t, k)\mathrm{e}^{\mathrm{i}kx} +$$

$$\frac{1}{2}\beta(t)(-4k^2)^n[a(t, k)\mathrm{e}^{-\mathrm{i}kx} + b(t, k)\mathrm{e}^{\mathrm{i}kx}] - [\beta(t)(-4k^2)^n x +$$

$$\alpha(t)(-4k^2)^{n+s-1}][-\mathrm{i}ka(t, k)\mathrm{e}^{-\mathrm{i}kx} + \mathrm{i}kb(t, k)\mathrm{e}^{\mathrm{i}kx}]$$

$$= \left[(-1)^n \frac{1}{2}\beta(t)(2k)^{2n} + (-1)^{n+s-1}\alpha(t)(2k)^{2(n+s-1)}\right][a(t, k)\mathrm{e}^{-\mathrm{i}kx} +$$

$$b(t, k)\mathrm{e}^{\mathrm{i}kx}] \qquad (3.2.37)$$

分别比较系数，给出

$$\frac{\mathrm{d}a(t, k)}{\mathrm{d}t} = 0, \ \frac{\mathrm{d}b(t, k)}{\mathrm{d}t} = \mathrm{i}k\alpha(t)(-4k^2)^{n+s-1}b \quad (3.2.38)$$

解此方程得到(3.2.26c).

§3.3 KdV 系统方程族的类孤子解

§3.3.1 KdV 系统方程族的类孤子解

当反射系数 $R(t,\ k(t)) = 0$ 时，(3.2.19)可写成

$$K(t,\ x,\ y) + \sum_{j=1}^{l} c_j^2(t) e^{-\kappa_j(t)(x+y)} + \sum_{j=1}^{l} c_j^2(t) e^{-\kappa_j(t)y} \int_x^{\infty} K(t,\ x,\ z) e^{-\kappa_j(t)z} dz = 0$$

$$(3.3.1)$$

这是一个退化核的积分方程. 设

$$K(t,\ x,\ y) = \sum_{m=1}^{l} c_m(t) h_m(t,\ x) e^{-\kappa_m(t)y} \qquad (3.3.2)$$

将其代入(3.3.1)，得到 $h_l(x)$ 的线性代数方程组

$$h_j(t,\ x) + \sum_{j=1}^{l} \frac{1}{\kappa_j(t) + \kappa_m(t)} c_j(t) c_m(t) e^{-(\kappa_j(t) + \kappa_m(t))x} h_m(t,\ x) = -c_j(t) e^{-\kappa_j(t)},$$

$$(j = 1,\ 2,\ \cdots,\ l) \qquad (3.3.3)$$

以 $D(t,\ x)$ 记 l 阶矩阵，其元素为

$$d_{ij} = \delta_{ij} + \frac{1}{\kappa_i(t) + \kappa_j(t)} c_i(t) c_j(t) e^{-(\kappa_i(t) + \kappa_j(t))x} \qquad (3.3.4)$$

则方程(3.3.3)的解为

$$h_j(t,\ x) = \frac{\det D_j(t,\ x)}{\det D(t,\ x)} \qquad (3.3.5)$$

式中 $D_j(t,\ x)$ 表示矩阵 $D(t,\ x)$ 中第 j 列元素由 $-c_m(t) e^{-\kappa_m(t)x}$，$(m=1,\ 2,\ \cdots,\ l)$ 依次替代后的矩阵. 将 $h_j(t,\ x)$ 代入(3.3.2)，得

$$K(t,\ x,\ x) = \frac{1}{\det D(t,\ x)} \sum_{j=1}^{l} c_m(t) \det D_m(t,\ x) e^{-\kappa_m x}$$

$$(3.3.6)$$

容易看出分子恰是行列式 $\det D(t, x)$ 对 x 的导数,所以

$$K(t, x, x) = \frac{\mathrm{d}}{\mathrm{d}x} \ln \det D(t, x) \tag{3.3.7}$$

从而 KdV 系统方程族无反射势的 l-类孤子解为

$$u(t, x) \doteq 2 \frac{\mathrm{d}^2}{\mathrm{d}x^2} \ln \det D(t, x) \tag{3.3.8}$$

§3.3.2   约化为等谱 KdV 方程族的解

当 $\beta(t) = 0$, $\alpha(t) = 1$ 时,方程族(3.1)变为等谱 KdV 方程族 (2.1.13a)而散射数据的演化关系式(3.2.26a, b)是

$$k_j(t) = k_j(0), \quad c_j(t) = c_j(0) \mathrm{e}^{-(2\kappa_j)^{2n+2s-2} \kappa_j t}, \quad (j = 1, 2, \cdots)$$

$$\tag{3.3.9a, b}$$

(1) 当 $n+s-1 = 1$, $l = 1$, 把(3.3.9)代入(3.3.8)得到等谱 KdV 方程(2.1.13b)的单孤子解

$$u = 2k^2 \operatorname{sech}^2 \frac{\left(2kx + 8k^3 t + \dfrac{1}{2} \ln \dfrac{c^2}{2k}\right)}{2} \tag{3.3.10a}$$

式中 c, k 是常数.

(2) 当 $n+s-1 = 1$, $l = 2$, 把(3.3.9)代入(3.3.8)得到等谱 KdV 方程(2.1.13b)的双孤子解

$$u = \frac{\begin{aligned}2c_1^2 k_1 \mathrm{e}^{\xi_1} + 2c_2^2 k_2 \mathrm{e}^{\xi_2} + \dfrac{2c_1^2 c_2^2}{k_1 k_2}(k_1 - k_2)^2 \mathrm{e}^{\xi_1 + \xi_2} + \\[2mm] \dfrac{c_1^4 c_2^2}{2k_1^2} \dfrac{k_2}{}\dfrac{(k_1 - k_2)^2}{(k_1 + k_2)^2} \mathrm{e}^{2\xi_1 + \xi_2} + c_1^2 c_2^4 \dfrac{k_1}{2k_2^2} \dfrac{(k_1 - k_2)^2}{(k_1 + k_2)^2} \mathrm{e}^{\xi_1 + 2\xi_2}\end{aligned}}{\left[1 + \dfrac{c_1^2}{2k_1} \mathrm{e}^{\xi_1} + \dfrac{c_2^2}{2k_2} \mathrm{e}^{\xi_2} + \dfrac{c_1^2 c_2^2}{4k_1 k_2} \dfrac{(k_1 - k_2)^2}{(k_1 + k_2)^2} \mathrm{e}^{\xi_1 + \xi_2}\right]^2}$$

$$\tag{3.3.10b}$$

其中 $\xi_i = -2k_i x - 8k_i^3 t$，$c_i$，$k_i$ 是常数，$i = 1$，2. 以上结果和[15]一致.

(3) 当 $n+s-1 = 2$，$l = 1$，把(3.3.9)代入(3.3.8)得到 KdV 方程(2.1.13c)的单孤子解

$$u = 2k^2 \operatorname{sech}^2\left(kx + 18k^5 t + \frac{1}{4}\ln\frac{c^2}{2k}\right) \qquad (3.3.11)$$

式中 c，k 是常数.

§3.3.3　约化为非等谱 KdV 方程族的解

当 $\beta(t) = 1$，$\alpha(t) = 0$ 时，方程族(3.1)变为非等谱 KdV 方程族(2.1.14a)而散射数据的演化关系式(3.2.26a, b)是

$$\kappa_j(t) = \frac{1}{(\kappa_j^{-2n}(0) - 2n \cdot 4^n t)^{\frac{1}{2n}}} \qquad (3.3.12a)$$

$$c_j(t) = c_j(0)\left(\frac{\kappa_j^{-2n}(0)}{\kappa_j^{-2n}(0) - 2n \cdot 4^n t}\right)^{\frac{2n+1}{4n}}, \quad (j = 1, 2, \cdots, l)$$

$$\qquad (3.3.12b)$$

(1) 当 $n = 1$，2，$l = 1$，把(3.3.12)代入(3.3.8)，非等谱 KdV 方程(2.1.14b, c)的单孤子解分别为

$$u(t, x) = \frac{1}{2(c - 2t)}\operatorname{sech}^2\left(\frac{1}{2}\ln(c - 2t) + \frac{x}{2\sqrt{c - 2t}}\right)$$

$$\qquad (3.3.13a)$$

$$u(t, x) = \frac{1}{2\sqrt{c - 4t}}\operatorname{sech}^2\left(\frac{1}{2}\ln(c - 4t) + \frac{x}{2\sqrt[4]{c - 4t}}\right)$$

$$\qquad (3.3.13b)$$

式中 c 为常数.

(2) 当 $n=1,2$，$l=2$，把(3.3.12)代入(3.3.8)，非等谱 KdV 方程(2.1.14b, c)的双孤子解分别为

$$u(t,x) = 2\frac{\frac{e^{\theta_1}}{c_1-2t} + \frac{e^{\theta_2}}{c_2-2t} + 2\left(\frac{1}{\sqrt{c_1-2t}} - \frac{1}{\sqrt{c_2-2t}}\right)^2 e^{\theta_1+\theta_2} + \frac{Ae^{\theta_1+2\theta_2}}{c_1-2t} + \frac{Ae^{2\theta_1+\theta_2}}{c_2-2t}}{(1+e^{\theta_1}+e^{\theta_2}+Ae^{\theta_1+\theta_2})^2}$$

$$(3.3.14a)$$

式中 c_1 和 c_2 为常数，$\theta_j = -\ln(c_1-2t) - \dfrac{x}{\sqrt{c_1-2t}}$，$j=1,2$，

$$A = \left(\frac{\sqrt{c_1+2t}-\sqrt{c_2+2t}}{\sqrt{c_1+2t}+\sqrt{c_2+2t}}\right)^2,$$

$$u(t,x) =$$
$$2\frac{\frac{e^{\theta_1}}{\sqrt{c_1-4t}} + \frac{e^{\theta_2}}{\sqrt{c_2-4t}} + 2\left(\frac{1}{\sqrt[4]{c_1-4t}} - \frac{1}{\sqrt[4]{c_2-4t}}\right)^2 e^{\theta_1+\theta_2} + \frac{Ae^{\theta_1+2\theta_2}}{\sqrt{c_1-4t}} + \frac{Ae^{2\theta_1+\theta_2}}{\sqrt{c_2-4t}}}{(1+e^{\theta_1}+e^{\theta_2}+Ae^{\theta_1+\theta_2})^2}$$

$$(3.3.14b)$$

式中 c_1 和 c_2 是常数，$\theta_j = -\ln(c_j-4t) - \dfrac{x}{\sqrt[4]{c_j-4t}}$，$j=1,2$，

$$A = \left(\frac{\sqrt[4]{c_1-4t}-\sqrt[4]{c_2-4t}}{\sqrt[4]{c_1-4t}+\sqrt[4]{c_2-4t}}\right)^2.$$

§3.3.4　约化为 τ 方程族的解

当 $\beta(t)=1$，$\alpha(t)=(2s-1)t$ 时，方程族(3.1)化为 τ 方程族 (2.1.15a)而散射数据的演化关系式(3.2.26a, b)是

$$\kappa_j(t) = \frac{1}{(\kappa_j^{-2n}(0) - 2n \cdot 4^n t)^{\frac{1}{2n}}} \qquad (3.3.15a)$$

51

$$c_j(t) = c_j(0)\left[\frac{\kappa_j^{-2n}(0)}{\kappa_j^{-2n}(0) - 2n \cdot 4^n t}\right]^{\frac{2n+1}{4n}} \Delta, \ (j = 1, 2, \cdots, l)$$

$$(3.3.15b)$$

$$\Delta = e^{\frac{(2s+1)4^{s-1}}{1-2s}t(k_j^{-2n}(0) - 2n4^n t)^{\frac{1-2s}{2n}} - \frac{(2s+1)4^{s-1-n}}{(1-2s)(1+2n-2s)}(k_j^{-2n}(0) - 2n4^n t)^{\frac{1+2n-2s}{2n}} + \frac{(2s+1)4^{s-1-n}}{(1-2s)(1+2n-2s)}k_j^{2s-2n-1}(0))}$$

特别地，当 $n = 0, s = 2, l = 1$ 时，τ 方程(2.1.15b)的单孤子解为

$$u = 2k^2 e^{2t} \text{sech}^2\left(kx e^t + \frac{4}{3}k^3 e^{3t}(3t-1) - \frac{4}{3}k^3 - \frac{1}{2}\ln\frac{c^2}{2k}\right)$$

$$(3.3.16)$$

式中 c, k 是常数.

§3.4 解的性质

§3.4.1 非等谱 KdV 方程解的性质

以非等谱 KdV 方程(2.1.14b)为例来考察非等谱情形下解的性质. 图 3(a)给出了非等谱 KdV 方程(2.1.14b)的单孤波解(3.3.13a)当 $c = 4$ 时的运动情况. 由图 3 可以看出波幅和波宽随时间变化，在 $t = 2$ 波幅趋向于 $+\infty$，即在 $t = 2$ 时，解出现奇异. 这由波幅的表达式 $\frac{1}{2(c-2t)}$ 容易看出. 同时波速随时间变化，事实上，波速为

$$\frac{dx}{dt} = -\frac{\ln(c-2t)}{\sqrt{c-2t}} - 2\frac{1}{\sqrt{c-2t}} \tag{3.4.1}$$

当 $t < -\frac{1}{2}e^{-1} + \frac{1}{2}$，$\frac{dx}{dt} < 0$；当 $t > -\frac{1}{2}e^{-1} + \frac{1}{2}$，$\frac{dx}{dt} > 0$，且速度变化很快，从速度图可以看出，速度由小于零到大于零，波形很快破裂，因

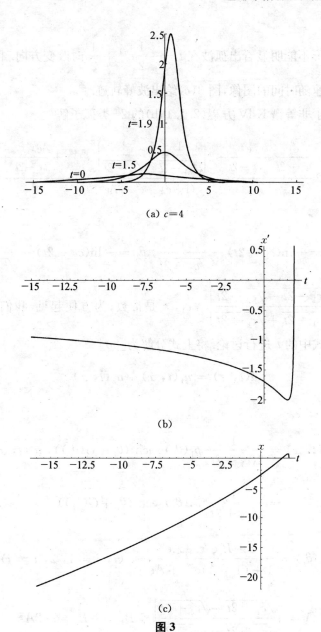

(a) $c=4$

(b)

(c)

图 3

此从图形不能明显看出孤波在 $t=-\dfrac{1}{2}\mathrm{e}^{-1}+\dfrac{1}{2}$ 时改变方向. 图 3(b)

给出波速随时间的图像, 图 3(c)给出波峰轨迹.

对于非等谱 KdV 方程(2.1.14b)的 2 -类孤子解

$$u(t, x)=2 \cdot \frac{\dfrac{\mathrm{e}^{\theta_1}}{c_1-2t}+\dfrac{\mathrm{e}^{\theta_2}}{c_2-2t}+2\left(\dfrac{1}{\sqrt{c_1-2t}}-\dfrac{1}{\sqrt{c_2-2t}}\right)^2\mathrm{e}^{\theta_1+\theta_2}+\dfrac{A\mathrm{e}^{\theta_1+2\theta_2}}{c_1-2t}+\dfrac{A\mathrm{e}^{2\theta_1+\theta_2}}{c_2-2t}}{(1+\mathrm{e}^{\theta_1}+\mathrm{e}^{\theta_2}+A\mathrm{e}^{\theta_1+\theta_2})^2}$$

$$(3.4.2)$$

其中 $\theta_1=-\ln(c_1-2t)-\dfrac{x}{\sqrt{c_1-2t}}$, $\theta_2=-\ln(c_2-2t)-\dfrac{x}{\sqrt{c_2-2t}}$,

$A=\left[\dfrac{\sqrt{c_1-2t}-\sqrt{c_2-2t}}{\sqrt{c_1-2t}+\sqrt{c_2-2t}}\right]^2$, c_1, c_2 是常数; 为方便起见, 我们以 $-t$

代替上式中的 t 进行讨论. 将上式分解为[80]

$$u(t, x)=u_1(t, x)+u_2(t, x) \qquad (3.4.3)$$

这里

$$u_1(t, x)=\dfrac{1}{2(c_1+2t)}p_1(\theta_2)\operatorname{sech}^2(\theta_1+G(\theta_2)), u_2(t, x)$$

$$=\dfrac{1}{2(c_2+2t)}p_2(\theta_1)\operatorname{sech}^2(\theta_2+G(\theta_1)) \qquad (3.4.4)$$

$$p_i(\theta_j)=\dfrac{1+B_j\mathrm{e}^{2\theta_j}+A\mathrm{e}^{4\theta_j}}{1+(1+A)\mathrm{e}^{2\theta_j}+A\mathrm{e}^{4\theta_j}}, (i, j=1, 2, i\neq j)$$

$$A=\left[\dfrac{\sqrt{c_1+2t}-\sqrt{c_2+2t}}{\sqrt{c_1+2t}+\sqrt{c_2+2t}}\right]^2, B_1=-B_2=-2A^{\frac{1}{2}}$$

$$G(\theta_j) = \frac{1}{2}\ln\left(\frac{1+Ae^{2\theta_j}}{1+e^{2\theta_j}}\right)$$

$$\theta_j = -\frac{1}{2}\left[\ln(c_j+2t)+\frac{x}{\sqrt{c_j+2t}}\right], \ (j=1,\ 2)$$

当 $t \to +\infty$ 时，$e^{\frac{-x}{\sqrt{c_j+2t}}}$ 很快地趋于 1，有

$$p_i(\theta_j) \approx \frac{(c_j+2t+\sqrt{A})^2}{(c_j+2t)^2+(c_j+2t)(1+A)+A}$$

$$G(\theta_j) \approx \frac{1}{2}\ln\frac{c_j+2t+A}{c_j+2t+1}, \ (t \to +\infty)$$

因此，对于固定的 θ_1，当 $t \to +\infty$，算得

$$u_1(t,\ x) \approx \frac{1}{2(c_1+2t)}\frac{(c_2+2t+\sqrt{A})^2}{(c_2+2t)^2+(c_2+2t)(1+A)+A}$$

$$\text{sech}^2\left(\theta_1+\frac{1}{2}\ln\frac{c_2+2t+A}{c_2+2t+1}\right) \tag{3.4.5}$$

对于固定的 θ_2，当 $t \to +\infty$ 时，又得

$$u_2(t,\ x) \approx \frac{1}{2(c_2+2t)}\frac{(c_1+2t+\sqrt{A})^2}{(c_1+2t)^2+(c_1+2t)(1+A)+A}$$

$$\text{sech}^2\left(\theta_2+\frac{1}{2}\ln\frac{c_1+2t+A}{c_1+2t+1}\right) \tag{3.4.6}$$

假定 $c_1 < c_2$，孤立波 u_2 早于孤立波 u_1 产生，当 u_2 向左运动和随后产生的 u_1 发生碰撞，碰撞后相互分离，但各自的振幅和相位都发生了改变. 图 4 给出 2-类孤子解相互作用到分离的情形（t 从 -1 到 ∞）.

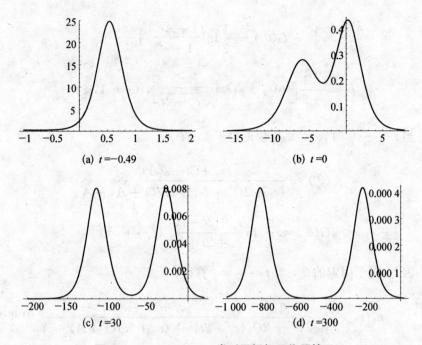

图 4 $c_1=1$, $c_2=3$, 2-类孤子解相互作用情况

§3.4.2 τ 方程解的性质

对于 τ 方程(2.1.15b)，由其单孤子解(3.3.16)知道，振幅 $2k^2 e^{2t}$ 随时间变化，当 $t \to -\infty$ 时，趋向于 $+\infty$，而波速为

$$\frac{dx}{dt} = -8k^2 t e^{2t} - \frac{4}{3}k^2 e^{2t} - \frac{4}{3}k^2 e^{-t} - \frac{1}{2}e^{-t}\ln\left(\frac{c^2}{2k}\right) \quad (3.4.7)$$

当 $t < \frac{1}{3}\text{LambertW}\left(-\dfrac{\left(8k^3+3\ln\dfrac{c^2}{2k}\right)e^{\frac{1}{2}}}{16k^3}\right) - \dfrac{1}{6}$ 时，速度大于零；

当 $t > \frac{1}{3}\text{LambertW}\left(-\dfrac{\left(8k^3+3\ln\dfrac{c^2}{2k}\right)e^{\frac{1}{2}}}{16k^3}\right) - \dfrac{1}{6}$ 时，速度小于零；

波在 $t = \dfrac{1}{3}\,\mathrm{LambertW}\left(-\dfrac{\left(8k^3 + 3\ln\dfrac{c^2}{2k}\right)\mathrm{e}^{\frac{1}{2}}}{16k^3}\right) - \dfrac{1}{6}$ 时改变传播方

向，这里 $\mathrm{LambertW}(x)$ 函数满足 $\mathrm{LambertW}(x) * \mathrm{e}^{\mathrm{LambertW}(x)} = x$.

如果取 $k = \dfrac{1}{2}$，$c = \mathrm{e}^{-1}$，则在 $t = -0.652\,996\,640\mathrm{e}^{-1}$ 时，波改变

传播方向.

图 5(a)给出了单类孤子解情形，图 5(b)给出了单类孤子解的速度变化

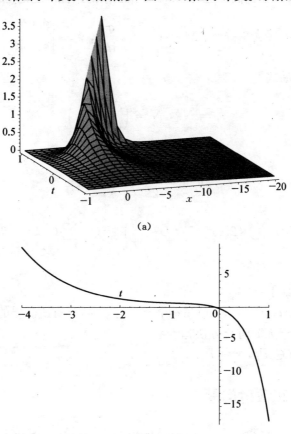

(a)

(b)

图 5

第四章 AKNS 系统方程族的 类孤子解

本章主要考察 AKNS 系统方程族(2.3.9)的类孤子解及其约化问题. 在第二章我们知道与(2.3.9)相联系的线性问题是

$$\phi_x = M\phi, \ M = \begin{pmatrix} -\mathrm{i}k & q \\ r & \mathrm{i}k \end{pmatrix}, \ \phi = \begin{pmatrix} \phi_1 \\ \phi_2 \end{pmatrix} \tag{4.1a}$$

和时间发展式

$$\phi_t = N\phi, \ N = \begin{pmatrix} A & B \\ C & -A \end{pmatrix} \tag{4.1b}$$

其中

$$k_t = -\frac{\mathrm{i}}{2}\beta(t)(2\mathrm{i}k)^n \tag{4.2}$$

$$\begin{pmatrix} B \\ C \end{pmatrix} = \alpha(t) \sum_{j=1}^{m+n-1} (2\mathrm{i}k)^{m+n-1-j} \widetilde{L}^{j-1} \begin{pmatrix} q \\ r \end{pmatrix} + \beta(t) \sum_{j=1}^{n} (2\mathrm{i}k)^{n-j} \widetilde{L}^{j-1} \begin{pmatrix} xq \\ xr \end{pmatrix} \tag{4.3}$$

假定如下:

(1) $q(t, x)$, $r(t, x)$在整个 x 轴有定义, 具有所需要的各阶导数且当 x 趋于无穷时充分快趋于零

(2) $q \in P_\mu$, $r \in P_\mu$.

§4.1 正散射问题

给定位势函数 q, r, 研究线性问题(4.1a)的特征值 k 以及特征

函数的性质称为正散射问题.

§4.1.1 特征函数的性质

在假定(1)、(2)下,线性问题(4.1a)有两组解$\{\phi(x, k), \bar{\phi}(x, k)\}$,$\{\psi(x, k), \bar{\psi}(x, k)\}$,它们是$x$的有界函数且具有渐进性质:

当$x \to +\infty$

$$\phi(x, k) \sim \begin{pmatrix} 0 \\ 1 \end{pmatrix} e^{ikx}, \qquad \bar{\phi}(x, k) \sim \begin{pmatrix} 1 \\ 0 \end{pmatrix} e^{-ikx}$$

$$\phi_x(x, k) \sim ik \begin{pmatrix} 0 \\ 1 \end{pmatrix} e^{ikx}, \qquad \bar{\phi}_x(x, k) \sim -ik \begin{pmatrix} 1 \\ 0 \end{pmatrix} e^{-ikx}$$

$$\phi_k(x, k) \sim ix \begin{pmatrix} 0 \\ 1 \end{pmatrix} e^{ikx}, \qquad \bar{\phi}_k(x, k) \sim -ix \begin{pmatrix} 1 \\ 0 \end{pmatrix} e^{-ikx}$$

$$(4.1.1a)$$

而当$x \to -\infty$

$$\psi(x, k) \sim \begin{pmatrix} 1 \\ 0 \end{pmatrix} e^{-ikx}, \qquad \bar{\psi}(x, k) \sim \begin{pmatrix} 0 \\ -1 \end{pmatrix} e^{ikx}$$

$$\psi_x(x, k) \sim -ik \begin{pmatrix} 1 \\ 0 \end{pmatrix} e^{-ikx}, \qquad \bar{\psi}_x(x, k) \sim ik \begin{pmatrix} 0 \\ -1 \end{pmatrix} e^{ikx}$$

$$\psi_k(x, k) \sim -ix \begin{pmatrix} 1 \\ 0 \end{pmatrix} e^{-ikx}, \qquad \bar{\psi}_k(x, k) \sim ix \begin{pmatrix} 0 \\ -1 \end{pmatrix} e^{ikx}$$

$$(4.1.1b)$$

其中$\phi(x, k)$和$\psi(x, k)$关于谱参数k在上半复平面解析且连续至实轴;$\bar{\phi}(x, k)$而$\bar{\psi}(x, k)$关于谱参数k在下半复平面解析且连续至实轴,并满足

当$k \to \infty$

$$\phi(x, k) \sim \begin{pmatrix} 0 \\ 1 \end{pmatrix} e^{ikx}, \qquad \bar{\phi}(x, k) \sim \begin{pmatrix} 1 \\ 0 \end{pmatrix} e^{-ikx}$$

$$\psi(x, k) \sim \begin{pmatrix} 1 \\ 0 \end{pmatrix} e^{-ikx}, \qquad \overline{\psi}(x, k) \sim \begin{pmatrix} 0 \\ -1 \end{pmatrix} e^{ikx} \qquad (4.1.2)$$

物理上称具有以上性质的解为 Jost 解.

事实上,求解线性问题(4.1a)等价与求解积分方程

$$\phi_1(x, k) = -\int_x^\infty e^{-ik(x-y)} q(y) \phi_2(y, k) \mathrm{d}y \qquad (4.1.3a)$$

$$\phi_2(x, k) = e^{ikx} - \int_x^\infty e^{ik(x-y)} r(y) \phi_1(y, k) \mathrm{d}y \qquad (4.1.3b)$$

$$\overline{\phi}_1(x, k) = e^{-ikx} - \int_x^\infty e^{-ik(x-y)} q(y) \overline{\phi}_2(y, k) \mathrm{d}y \qquad (4.1.3c)$$

$$\overline{\phi}_2(x, k) = -\int_x^\infty e^{ik(x-y)} r(y) \overline{\phi}_1(y, k) \mathrm{d}y \qquad (4.1.3d)$$

与

$$\psi_1(x, k) = e^{-ikx} + \int_{-\infty}^x e^{-ik(x-y)} q(y) \psi_2(y, k) \mathrm{d}y \qquad (4.1.4a)$$

$$\psi_2(x, k) = \int_{-\infty}^x e^{ik(x-y)} r(y) \psi_1(y, k) \mathrm{d}y \qquad (4.1.4b)$$

$$\overline{\psi}_1(x, k) = \int_{-\infty}^x e^{-ik(x-y)} q(y) \overline{\psi}_2(y, k) \mathrm{d}y \qquad (4.1.4c)$$

$$\overline{\psi}_2(x, k) = -e^{ikx} + \int_{-\infty}^x e^{ik(x-y)} r(y) \overline{\psi}_1(y, k) \mathrm{d}y \qquad (4.1.4d)$$

为了说明这些积分方程组解的存在性,例如(4.1.3a, b),作变换

$$f(x, k) = e^{-ikx} \phi(x, k) \qquad (4.1.5)$$

则此方程组化为

$$f_1(x, k) = -\int_x^\infty e^{-2ik(x-y)} q(y) f_2(y, k) \mathrm{d}y \qquad (4.1.6a)$$

$$f_2(x, k) = 1 - \int_x^\infty r(y) f_1(y, k) \mathrm{d}y \qquad (4.1.6\mathrm{b})$$

从而可得

$$f_2(x, k) = 1 + \int_x^\infty r(y) \left(\int_y^\infty \mathrm{e}^{-2ik(y-z)} q(z) f_2(z, k) \mathrm{d}z \right) \mathrm{d}y$$

$$(4.1.7)$$

按逐次逼近法构造函数列

$$f_2^{(0)}(x, k) = 1 \qquad (4.1.8\mathrm{a})$$

$$f_2^{(j)}(x, k) = \int_x^\infty r(y) \left(\int_y^\infty \mathrm{e}^{-2ik(y-z)} q(z) f_2^{(j-1)}(z, k) \mathrm{d}z \right) \mathrm{d}y \ (j = 1, 2, \cdots)$$

$$(4.1.8\mathrm{b})$$

则当 $\mathrm{Im}\, k \geqslant 0$ 时有估计式

$$|f_2^{(j)}(x, k)| \leqslant \frac{1}{j!^2} (Q_0(x) R_0(x))^j \qquad (4.1.9)$$

其中

$$Q_l(x) = \int_x^\infty |y^l q(y)| \, \mathrm{d}y, \ R_l(x) = \int_x^\infty |y^l r(y)| \, \mathrm{d}y$$

$$(4.1.10)$$

由此可见函数项级数

$$f_2^{(0)}(x, k) + f_2^{(1)}(x, k) + \cdots + f_2^{(j)}(x, k) + \cdots \qquad (4.1.11)$$

当 $\mathrm{Im}\, k \geqslant 0$ 时在整个 x 轴上绝对一致收敛, 其和即为 $f_2(x, k)$ 且成立不等式

$$|f_2(x, k)| \leqslant I_0(2\sqrt{Q_0(x) R_0(x)}) \qquad (4.1.12)$$

这里 $I_0(x)$ 表示零阶修正 Bessel 函数. $f_2(x, k)$ 确定之后, 可由

(4.1.6a)计算 $f_1(x, k)$.

若积分方程(4.1.7)还存在另一个解 $\tilde{f}_2(x, k)$，则差 $f_2(x, k) - \tilde{f}_2(x, k)$ 满足不等式

$$| f_2(x,k) - \tilde{f}_2(x,k) | \leqslant \int_x^\infty | r(y) | \left(\int_y^\infty | q(z) | | f_2(z,k) - \tilde{f}_2(z,k) | \mathrm{d}z \right) \mathrm{d}y \tag{4.1.13}$$

将上式右端对 z 的积分下限 y 代之以 x，积分值不会减小，故得

$$| f_2(x,k) - \tilde{f}_2(x,k) | \leqslant R_0(x) \int_x^\infty | q(z) | | f_2(z,k) - \tilde{f}_2(z,k) | \mathrm{d}z \tag{4.1.14}$$

但 $R_0(x) \leqslant R_0 = \int_{-\infty}^\infty | r(x) | \mathrm{d}x$，并令

$$F(x, k) = | f_2(x,k) - \tilde{f}_2(x,k) | \tag{4.1.15a}$$

$$G(x, k) = R_0 \int_x^\infty | q(z) | | F(z,k) | \mathrm{d}z \tag{4.1.15b}$$

则有

$$F(x, k) \leqslant G(x, k) \tag{4.1.16}$$

将(4.1.15b)对 x 求导数后并利用此不等式得

$$G_x(x, k) \geqslant - R_0 | q(x) | G(x, k) \tag{4.1.17}$$

或写成

$$(e^{-R_0 \int_x^\infty | q(y) | \mathrm{d}y} G(x, k))_x \geqslant 0 \tag{4.1.18}$$

可见括号内的函数是非负单增的. 但当 x 趋于正无穷时，它趋于零，所以 $G(x, k)$ 恒等于零. 这就是说

$$f_2(x, k) = \tilde{f}_2(x, k) \tag{4.1.19}$$

回到变换(4.1.5)，得到满足线性问题(4.1a)及渐近条件(4.1.1a)的 Jost 函数 $\phi(x, k)$ 在域 $\mathrm{Im}k \geqslant 0 - \infty < x < \infty$ 上存在唯一，且关于 x 及 k 是连续的.

类似的推理也说明满足线性问题(4.1a)及其他渐近条件的 Jost 函数的存在唯一性以及 Jost 函数关于 x, k 的可微性.

定义 4.1 $\phi(x, k)$ 和 $\psi(x, k)$ 的 Wronski 行列式 $W(\phi(x, k), \psi(x, k))$ 为：

$$W(\phi(x, k), \psi(x, k)) = \phi_1(x, k)\psi_2(x, k) - \phi_2(x, k)\psi_1(x, k)$$

$$(4.1.20)$$

根据(4.1a)和 Jost 解的渐近性质(4.1.1a, b)，有

$$W(\overline{\phi}(x, k), \phi(x, k)) = 1, \quad W(\psi(x, k), \overline{\psi}(x, k)) = -1$$

$$(4.1.21a, b)$$

§4.1.2 反射系数和穿透系数

对于每一个实数 k, Jost 函数组 $\{\phi(x, k), \overline{\phi}(x, k)\}$ 和 $\{\psi(x, k), \overline{\psi}(x, k)\}$ 分别构成线性问题(4.1a)的基本解组，它们之间存在线性关系：

$$\psi(x, k) = a(k)\overline{\phi}(x, k) + b(k)\phi(x, k) \qquad (4.1.22a)$$

$$\overline{\psi}(x, k) = -\bar{a}(k)\phi(x, k) + \bar{b}(k)\overline{\phi}(x, k) \qquad (4.1.22b)$$

由(4.1.21a)，(4.1.22)容易算得相关系数

$$a(k) = W(\psi(x, k), \phi(x, k)), \quad \bar{a}(k) = W(\overline{\psi}(x, k), \overline{\phi}(x, k))$$

$$(4.1.23a)$$

$$b(k) = W(\overline{\phi}(x, k), \psi(x, k)), \quad \bar{b}(k) = W(\overline{\psi}(x, k), \phi(x, k))$$

$$(4.1.23b)$$

由 $W(\psi(x, k), \overline{\psi}(x, k)) = W(a(k)\overline{\phi} + b(k)\phi, -\bar{a}(k)\phi + \bar{b}(k)\overline{\phi}) =$

$-a\bar{a} - b\bar{b} = -1$，得

$$a\bar{a} + b\bar{b} = 1 \qquad (4.1.24)$$

因此对任意的实数 k，$a(k)$，$\bar{a}(k)$ 必不等于零. (4.1.22a, b) 又可写成

$$T(k)\psi(x, k) = \bar{\phi}(x, k) + R(k)\phi(x, k) \qquad (4.1.25a)$$

$$\bar{T}(k)\bar{\psi}(x, k) = -\phi(x, k) + \bar{R}(k)\bar{\phi}(x, k) \qquad (4.1.25b)$$

其中 $T(k) = \dfrac{1}{a(k)}$，$\bar{T}(k) = \dfrac{1}{\bar{a}(k)}$ 与 $R(k) = \dfrac{b(k)}{a(k)}$，$\bar{R}(k) = \dfrac{\bar{b}(k)}{\bar{a}(k)}$ 分别称为穿透系数与反射系数.

§4.1.3　谱的分布

引理 4.1　$a(k)$ 在上半 k 平面解析，$\bar{a}(k)$ 在下半 k 平面解析；$b(k)$ 和 $\bar{b}(k)$ 仅在实轴有定义，且

$$a(k) = 1 + \int_{-\infty}^{\infty} e^{iky} q(y) \psi_2(y, k) dy \qquad (4.1.26a)$$

$$\bar{a}(k) = 1 - \int_{-\infty}^{\infty} e^{-iky} r(y) \bar{\psi}_1(y, k) dy \qquad (4.1.26b)$$

$$b(k) = \int_{-\infty}^{\infty} e^{-iky} r(y) \psi_1(y, k) dy \qquad (4.1.26c)$$

$$\bar{b}(k) = \int_{-\infty}^{\infty} e^{-iky} q(y) \bar{\psi}_2(y, k) dy \qquad (4.1.26d)$$

证明： 首先，根据 Jost 函数的解析性质，由 (4.1.23a, b) 知 $a(k)$ 在上半 k 平面解析，$\bar{a}(k)$ 在下半 k 平面解析；$b(k)$ 和 $\bar{b}(k)$ 仅在实轴有定义.

其次改写 (4.1.22a)，并令

$$\Delta = \psi(x, k) - a(k)\bar{\phi}(x, k) = b(k)\phi(x, k), \quad \Delta = (\Delta_1, \Delta_2)^{\mathrm{T}} \tag{4.1.27}$$

将(4.1.3)，(4.1.4a，b)代入上式,算得

$$\Delta_1 = e^{-ikx} - a(k)e^{-ikx} + \int_{-\infty}^{\infty} e^{-ik(x-y)}q(y)\psi_2(y,k)dy -$$

$$\int_x^{\infty} e^{-ik(x-y)}q(y)\Delta_2(y,k)dy$$

$$\Delta_2 = \int_{-\infty}^{\infty} e^{ik(x-y)}r(y)\psi_1(y,k)dy - \int_x^{\infty} e^{ik(x-y)}r(y)\Delta_1(y,k)dy$$

又

$$\Delta_1 = -\int_x^{\infty} e^{-ik(x-y)}q(y)\Delta_2(y,k)dy$$

$$\Delta_2 = b(k)e^{ikx} - \int_x^{\infty} e^{ik(x-y)}r(y)\Delta_1(y,k)dy$$

因此

$$a(k) = 1 + \int_{-\infty}^{\infty} e^{iky}q(y)\psi_2(y,k)dy$$

$$b(k) = \int_{-\infty}^{\infty} e^{-iky}r(y)\psi_1(y,k)dy$$

类似计算可给出$\bar{a}(k)$，$\bar{b}(k)$.

引理 4.2 函数$a(k)(\bar{a}(k))$在上半k平面(下半k平面)只有有限个零点k_1，k_2，$\cdots$，$k_l(\bar{k}_1,\bar{k}_2,\cdots,\bar{k}_{\bar{l}})$.

证明:由(4.1.26a，b)知,当$k\to\infty$时,$a(k)\sim 1$,$\bar{a}(k)\sim 1$,利用解析函数的性质,知$a(k)(\bar{a}(k))$的零点必定是有限个.

如果k_j是$a(k)$的零点,即

$$W(\phi(x,k_j),\psi(x,k_j)) = 0$$

则$\phi(x,k_j)$和$\psi(x,k_j)$是线性相关的,存在常数b_j使得

$$\psi(x,k_j) = b_j\phi(x,k_j) \tag{4.1.28}$$

由此推知当x趋于正无穷或负无穷时,$\psi(x,k_j)$均呈指数衰减而在整

个数轴上平方可积,所以它是线性问题(4.1a)的特征函数,k_j 是相应的特征值.

假设 $a(k)(\bar{a}(k))$ 在 k 的上(下)半平面的根是单重根,(4.1a)两端对 k 微商得

$$(\partial + ik)\phi_{1,k}(x, k) = -i\phi_1(x, k) + q(x)\phi_{2,k}(x, k)$$

(4.1.29a)

$$(\partial - ik)\phi_{2,k}(x, k) = i\phi_2(x, k) + r(x)\phi_{1,k}(x, k)$$

(4.1.29b)

以 $\psi_2(x, k)$ 乘(4.1.29a), $-\psi_1(x, k)$ 乘(4.1.29b), 再以 $-\phi_{2,k}(x, k)$, $\phi_{1,k}(x, k)$ 分别乘 $\psi(x, k)$ 满足的谱方程(4.1a), 然后将所得的四等式相加, 有

$$\frac{d}{dx}W(\psi(x, k), \phi_k(x, k)) = i(\phi_1(x, k)\psi_2(x, k) + \phi_2(x, k)\psi_1(x, k))$$

(4.1.30a)

类似地,有

$$\frac{d}{dx}W(\psi_k(x, k), \phi(x, k)) = -i(\phi_1(x, k)\psi_2(x, k) + \phi_2(x, k)\psi_1(x, k))$$

(4.1.30b)

从 x 至 l 积分等式(4.1.30a),从 $-l$ 至 x 积分等式(4.1.30b),并将所得结果相减后代入 $a(k)$ 的零点 k_j,再令 l 趋于无穷给出

$$\frac{d}{dk}W(\psi(x, k_j), \phi(x, k_j)) = -2ib_j \int_{-\infty}^{\infty} \phi_1(x, k_j)\phi_2(x, k_j)dx$$

(4.1.31)

这里利用线性相关式(4.1.28).但等式(4.1.31)的左端恰是 $a(k)$ 对 k 的微商在 k_j 的值,所以

$$2\int_{-\infty}^{\infty}\phi_1(x,k_j)\phi_2(x,k_j)\mathrm{d}x=\frac{\mathrm{i}a_k(k_j)}{b_j} \qquad (4.1.32)$$

则必存在常数 c_j，使得

$$2\int_{-\infty}^{\infty}c_j^2\phi_1(x,k_j)\phi_2(x,k_j)\mathrm{d}x=1 \qquad (4.1.33)$$

称此常数为特征函数 $\phi(x,k_j)$ 的归一化因子，而 $c_j\phi(x,k_j)$ 称为归一化特征函数. 从(4.1.33)可见

$$c_j^2=-\frac{\mathrm{i}b_j}{a_k(k_j)} \qquad (4.1.34a)$$

类似地，我们有

$$\bar{c}_j^2=-\frac{\mathrm{i}\bar{b}_j}{\bar{a}_k(\bar{k}_j)} \qquad (4.1.34b)$$

定义 4.2　如果 k_j 和 $\bar{k}_j$ 分别是 $a(k)$ 和 $\bar{a}(k)$ 的单重根，存在常数 c_j 和 $\bar{c}_j$ 使得

$$2\int_{-\infty}^{\infty}c_j^2\phi_1(x,k_j)\phi_2(x,k_j)\mathrm{d}x=1,\ 2\int_{-\infty}^{\infty}\bar{c}_j^2\bar{\phi}_1(x,\bar{k}_j)\bar{\phi}_2(x,\bar{k}_j)\mathrm{d}x=1$$
$$(4.1.35)$$

称 c_j 和 $\bar{c}_j$ 是特征函数 $\phi(x,k_j)$ 和 $\bar{\phi}(x,\bar{k}_j)$ 的归一化常数；$c_j\phi(x,k_j)$ 和 $\bar{c}_j\bar{\phi}(x,\bar{k}_j)$ 为归一化特征函数.

　　线性问题(4.1a)除了特征函数平方可积的离散谱 k_j 和 $\bar{k}_j$，还有特征函数不能归一化的连续谱，连续谱遍布整个实轴.

定义 4.3　称集合

$$\{k(\mathrm{Im}\,k=0),\,R(k)=\frac{b(k)}{a(k)},\,\bar{R}(k)=\frac{\bar{b}(k)}{\bar{a}(k)};$$

$$k_j(\mathrm{Im}\,k_j>0),\,\bar{k}_m(\mathrm{Im}\,\bar{k}_m<0),\,c_j,\,\bar{c}_m,\,j=1,2,\cdots,l,\,m=1,2,\cdots,\bar{l}\,\}$$
$$(4.1.36)$$

为线性问题(4.1a)的散射数据.

§4.2 反散射问题

在已知散射数据的基础上,要求恢复线性问题(4.1a)的位势 $q(x)$, $r(x)$,即为 AKNS 系统的方程族的反散射问题.

§4.2.1 平移变换与 GLM 积分方程

一、平移变换

求线性问题(4.1a)满足渐近条件(4.1.1a, b)的解等价于求解积分方程(4.1.3), (4.1.4),当 $q = r = 0$ 时,积分方程(4.1.3), (4.1.4)有解

$$\phi(x, k) = \begin{pmatrix} 0 \\ 1 \end{pmatrix} e^{ikx}, \qquad \bar{\phi}(x, k) = \begin{pmatrix} 1 \\ 0 \end{pmatrix} e^{-ikx}$$

$$(4.2.1a, b)$$

$$\psi(x, k) = \begin{pmatrix} 1 \\ 0 \end{pmatrix} e^{-ikx}, \qquad \bar{\psi}(x, k) = \begin{pmatrix} 0 \\ -1 \end{pmatrix} e^{ikx}$$

$$(4.2.1c, d)$$

则存在唯一的可微向量函数

$$K(x, y) = (K_1(x, y), K_2(x, y))^T$$

$$\bar{K}(x, y) = (\bar{K}_1(x, y), \bar{K}_2(x, y))^T (x < y)$$

$$J(x, y) = (J_1(x, y), J_2(x, y))^T$$

$$\bar{J}(x, y) = (\bar{J}_1(x, y), \bar{J}_2(x, y))^T (x > y)$$

将(4.2.1)变为(4.1a)的解. 即

$$\phi(x, k) \doteq \begin{pmatrix} 0 \\ 1 \end{pmatrix} e^{ikx} + \int_x^\infty K(x, y) e^{iky} dy \qquad (4.2.2a)$$

$$\bar{\phi}(x,\ k) = \binom{1}{0} e^{-ikx} + \int_x^\infty \overline{K}(x,\ y) e^{-iky}\,dy \qquad (4.2.2b)$$

$$\psi(x,\ k) = \binom{1}{0} e^{-ikx} + \int_{-\infty}^x J(x,\ y) e^{-iky}\,dy \qquad (4.2.2c)$$

$$\bar{\psi}(x,\ k) = \binom{0}{-1} e^{ikx} + \int_{-\infty}^x \overline{J}(x,\ y) e^{iky}\,dy \qquad (4.2.2d)$$

事实上,将(4.2.2a)代入方程(4.1.3a, b),给出 $K(x,\ y)$ 所满足的积分方程

$$K_1(x,\ y) = -\frac{1}{2} q\Big(\frac{x+y}{2}\Big) - \int_x^{\frac{x+y}{2}} q(s) K_2(s,\ x+y-s)\,ds$$

$$(4.2.3a)$$

$$K_2(x,\ y) = -\int_x^\infty r(s) K_1(s,\ -x+y+s)\,ds \qquad (4.2.3b)$$

从中消去 $K_2(x,\ y)$,得

$$K_1(x,\ y) = -\frac{1}{2} q\Big(\frac{x+y}{2}\Big) + \int_x^{\frac{x+y}{2}} \int_x^\infty q(s) r(z) K_1(z,\ x+y-2s+z)\,dz\,ds$$

$$(4.2.4)$$

当 $y = x$ 时,特别有,

$$K_1(x,\ x) = -\frac{1}{2} q(x) \qquad (4.2.5a)$$

$$K_2(x,\ x) = \frac{1}{2} \int_x^\infty q(s) r(s)\,ds \qquad (4.2.5b)$$

按逐次逼近法作函数列

$$K_1^{(0)}(x,\ y) = -\frac{1}{2} q\Big(\frac{x+y}{2}\Big) \qquad (4.2.6a)$$

$$K_1^{(j)}(x, y) = \int_x^{\frac{x+y}{2}} \int_s^\infty q(s)r(z)K_1^{(j-1)}(z, x+y-2s+z)\mathrm{d}z\mathrm{d}s \ (j = 1, 2, \cdots)$$

$$(4.2.6\mathrm{b})$$

以 q_0 记 $|q(x)|$ 在整个 x 轴上的最大值，则列中每个函数有估计式

$$| K_1^{(j)}(x, y) | \leqslant \frac{q_0}{2 \cdot j!^2}(Q_0(x)R_0(x))^j \ (j = 0, 1, 2, \cdots)$$

$$(4.2.7)$$

其中 $Q_0(x)$ 与 $R_0(x)$ 是积分 (4.1.10) $(l = 0)$. 于是级数

$$K_1^{(0)}(x, y) + K_1^{(1)}(x, y) + \cdots + K_1^{(j)}(x, y) + \cdots \quad (4.2.8\mathrm{a})$$

在 $y \geqslant x$ 的区域上绝对一致收敛，其和即为积分方程(4.2.4)的解 $K_1(x, y)$，且满足不等式

$$K_1(x, y) \leqslant \frac{q_0}{2} I_0(\sqrt{2Q_0(x)R_0(x)}) \qquad (4.2.8\mathrm{b})$$

若此方程还存在另一个解 $\widetilde{K}_1(x, y)$，令

$$F(x, y) = | K_1(x, y) - \widetilde{K}_1(x, y) | \qquad (4.2.9\mathrm{a})$$

$$G(x, y) = q_0 \int_x^{\frac{x+y}{2}} \int_s^\infty | r(z) | F(z, x+y-2s+z)\mathrm{d}z\mathrm{d}s$$

$$(4.2.9\mathrm{b})$$

于是有

$$F(x, y) \leqslant G(x, y) \qquad (4.2.10)$$

作自变数代换 $x' = \frac{1}{2}(x+y)$，$y' = \frac{1}{2}(x-y)$，积分(4.2.9b)化为

$$G(x'+y', x'-y') = q_0 \int_{x'+y'}^{x'} \int_s^\infty | r(z) | F(z, 2x'-2s+z)\mathrm{d}z\mathrm{d}s$$

$$(4.2.11)$$

再作积分变数代换 $s=s'+x'$，$z=s'+z'$，若仍记 $G(x'+y'，x'-y')$ 为 $G(x'，y')$，最后得

$$G(x'，y')=q_0\int_{y'}^0\Big(\int_{x'}^\infty\mid r(s'+z')\mid F(s'+z'，-s'+z')\mathrm{d}z'\Big)\mathrm{d}s'$$

$$(4.2.12)$$

将此式对 y' 微商有

$$G_{y'}(x'，y')=-q_0\int_{x'}^\infty\mid r(y'+z')\mid F(y'+z'，-y'+z')\mathrm{d}z'$$

$$(4.2.13)$$

右端积分中以 $G(z'，y')$ 替代 $F(y'+z'，-y'+z')$ 不减小，所以

$$G_{y'}(x'，y')\geqslant-q_0\int_{x'}^\infty\mid r(y'+z')\mid G(z'，y')\mathrm{d}z'$$

$$(4.2.14)$$

当 $z'\geqslant x'$ 时 $G(z'，y')\leqslant G(x'，y')$，此不等式简化成

$$G_{y'}(x'，y')\geqslant-q_0\Big(\int_{x'}^\infty\mid r(y'+z')\mid\mathrm{d}z'\Big)G(x'，y')$$

$$(4.2.15)$$

从而推得

$$(\mathrm{e}^{q_0\int_{x'+y'}^\infty(x'+y'-s)\mid r(s)\mid\mathrm{d}s}G(x'，y'))_{y'}\geqslant0 \qquad(4.2.16)$$

这说明括号中的函数是变量 y' 的单增函数. 但当 $y'=0$ 时 $G(x'，y')$ 的值为零，故在 $y'\leqslant0$（$x\leqslant y$）的域上，非负函数 $G(x'，y')$ 恒为零. 就是

$$K_1(x，y)=\widetilde{K}_1(x，y) \qquad(4.2.17)$$

二、GLM 积分方程

平移变换中的函数 $K(x，y)$ 与 $\overline{K}(x，y)$ 满足 GLM 积分方程，而此方程的核可通过线性问题(4.1a)的离散谱，归一化因子与连续

谱，反射系数这些散射数据明显地表达出来. 我们有如下定理.

定理 4.1 设给定线性问题(4.1a)的两组散射数据

$$\{\mathrm{Im}\, k = 0,\, R(k),\, k_j,\, c_j,\, j = 1, 2, \cdots, l\} \quad (4.2.18a)$$

$$\{\mathrm{Im}\, \overline{k} = 0,\, \overline{R}(k),\, \overline{k}_j,\, \overline{c}_j,\, j = 1, 2, \cdots, \overline{l}\} \quad (4.2.18b)$$

并令

$$F_c(x) = \frac{1}{2\pi} \int_{-\infty}^{\infty} R(k) \mathrm{e}^{\mathrm{i}kx} \mathrm{d}k,\ F_d(x) = \sum_{j=1}^{l} c_j^2 \mathrm{e}^{\mathrm{i}k_j x} \quad (4.2.19a)$$

$$\overline{F}_c(x) = \frac{1}{2\pi} \int_{-\infty}^{\infty} \overline{R}(k) \mathrm{e}^{-\mathrm{i}kx} \mathrm{d}k,\ \overline{F}_d(x) = -\sum_{j=1}^{\overline{l}} \overline{c}_j^{\,2} \mathrm{e}^{-\mathrm{i}\overline{k}_j x}$$

$$(4.2.19b)$$

$$F(x) = F_c(x) + F_d(x), \qquad \overline{F}(x) = \overline{F}_c(x) + \overline{F}_d(x)$$

$$(4.2.19c)$$

则平移变换的向量函数 $K(x, y)$ 与 $\overline{K}(x, y)$ 满足方程

$$\overline{K}(x, y) + \begin{pmatrix} 0 \\ 1 \end{pmatrix} F(x+y) + \int_x^{\infty} K(x, z) F(z+y) \mathrm{d}z = 0$$

$$(4.2.20a)$$

$$K(x, y) - \begin{pmatrix} 1 \\ 0 \end{pmatrix} \overline{F}(x+y) - \int_x^{\infty} \overline{K}(x, z) \overline{F}(z+y) \mathrm{d}z = 0$$

$$(4.2.20b)$$

证：从 Jost 函数的关系式(4.1.25a)的两端同减 $\begin{pmatrix} 1 \\ 0 \end{pmatrix} \mathrm{e}^{-\mathrm{i}kx}$，再对 k 作 Fourier 变换得

$$\int_{-\infty}^{\infty} \left(T(k)\psi(x, k) - \begin{pmatrix} 1 \\ 0 \end{pmatrix} \mathrm{e}^{-\mathrm{i}kx} \right) \mathrm{e}^{\mathrm{i}ky} \mathrm{d}k$$

$$= \int_{-\infty}^{\infty} \left(\not\phi(x, k) - \binom{1}{0} e^{-ikx} \right) e^{iky} dk + \int_{-\infty}^{\infty} R(k) \not\phi(x, k) e^{iky} dk$$

$$(4.2.21)$$

左端被积函数在 k 的上半平面内已设仅存在有限个简单极点 k_j，并且当 k 趋于无穷时为零，依据留数定理与 Jordan 引理，这积分的值为

$$2\pi i \sum_{j=1}^{l} \frac{\psi(x, k_j)}{a_k(k_j)} e^{ik_j y} \qquad (4.2.22)$$

应用等式 (4.1.28)，(4.1.34a) 与 $k = k_j (j = 1, 2, \cdots, l)$ 的平移变换 (4.2.2a)，改写它为

$$- 2\pi \binom{0}{1} \sum_{j=1}^{l} c_j^2 e^{ik_j(x+y)} - 2\pi \int_x^{\infty} K(x, z) \sum_{j=1}^{l} c_j^2 e^{ik_j(z+y)} dz$$

$$(4.2.23)$$

其次，(4.2.21) 右端的第一项积分在代入 (4.2.2b) 后并借助 Fourier 变换反演公式后给出 $2\pi \overline{K}(x, y)$. 而第二项积分在代入 (4.2.2a) 可得

$$\binom{0}{1} \int_{-\infty}^{\infty} R(k) e^{ik(x+y)} dk + \int_x^{\infty} K(x, z) \left(\int_{-\infty}^{\infty} R(k) e^{ik(z+y)} dk \right) dz$$

$$(4.2.24)$$

由假设条件 (4.2.19)，于是整个等式 (4.2.21) 化为

$$- \binom{0}{1} F_d(x+y) - \int_x^{\infty} K(x, z) F_d(z+y) dz$$

$$= \overline{K}(x, y) + \binom{0}{1} F_c(x+y) + \int_x^{\infty} K(x, z) F_c(z+y) dz$$

$$(4.2.25)$$

此即积分方程 (4.2.20a).

如果对函数关系式(4.1.25b)进行类似的分析则可导出方程
(4.2.20b).通常称(4.2.20)为 AKNS 系统方程族的 GLM 积分方程
组.一旦解出函数 $K(x, y)$ 后,位势 $q(x)$ 与 $r(x)$ 即可借助公式
(4.2.5)进行恢复.同样的方法可引出平移变换中向量函数 $J(x, y)$
与 $\bar{J}(x, y)$ 在所满足的积分方程.

§4.2.2　散射数据随时间的演化规律

引理 4.3　假设 $\phi(x, k)$ 是(4.1a)的一个解,M 和 N 满足零曲
率方程 $M_t - N_x = [N, M]$,则

$$P(x, k) = \phi_t(x, k) - N\phi(x, k) \qquad (4.2.26)$$

也是(4.1a)的解.

引理 4.4　$\tilde{L}^* = -\sigma\partial + 2\begin{pmatrix} -r \\ q \end{pmatrix}\partial^{-1}(q, r)$ 是 $\tilde{L} = \sigma L\sigma$ 的共轭
算子.

引理 4.5　如果 $\phi(x, k) = (\phi_1(x, k), \phi_2(x, k))^T$ 是线性问题
(4.1a)以 k 为特征值的特征函数,则 $(\phi_2^2(x, k), \phi_1^2(x, k))^T$ 是 $\tilde{L}^*$ 以
$2ik$ 为特征值的特征函数.

定理 4.2　线性问题(4.1a)的散射数据 $\{k_j(t), \bar{k}_j(t), c_j(t), \bar{c}_j(t),$
$R(t, k), \bar{R}(t, k)\}$ 服从演化规律

$$k_j(t) = \left[k_j^{1-n}(0) + (1-n)(2i)^{n-1}\int_0^t \beta(s)\mathrm{d}s \right]^{\frac{1}{1-n}}$$

$$(4.2.27a)$$

$$\bar{k}_h(t) = \left[\bar{k}_h^{1-n}(0) + (1-n)(2i)^{n-1}\int_0^t \beta(s)\mathrm{d}s \right]^{\frac{1}{1-n}}$$

$$(4.2.27b)$$

$$c_j(t) = c_j(0)\mathrm{e}^{\frac{1}{2}\left[\int_0^t (\alpha(s)(2ik_j(s))^{m+n-1}+n\beta(s)(2ik_j(s))^{n-1})\mathrm{d}s \right]}$$

$$(4.2.27c)$$

$$\overline{c}_h(t) = \overline{c}_h(0) e^{\frac{1}{2}\left[\int_0^t (\alpha(s)(2i\overline{k}_h(s))^{m+n-1} + n\beta(s)(2i\overline{k}_h(s))^{n-1})ds\right]}$$

$$(4.2.27d)$$

$$R(t, k) = R(0, k(0)) e^{\int_0^t \alpha(s)(2ik(s))^{m+n-1}ds} \qquad (4.2.27e)$$

$$\overline{R}(t, k) = \overline{R}(0, k(0)) e^{\int_0^t \alpha(s)(2i\overline{k}(s))^{m+n-1}ds} \qquad (4.2.27f)$$

其中 $j = 1, 2, \cdots, l$ 和 $h = 1, 2, \cdots, \overline{l}$；$k_j(0)$, $\overline{k}_h(0)$, $c_j(0)$, $\overline{c}_h(0)$, $R(0, k(0))$ 与 $\overline{R}(0, k(0))$ 是(4.1a)在 $(q(t, x), r(t, x))^T = (q(0, x), r(0, x))^T$ 时的散射数据.

特别地，当 $n = 1$ 时，(4.2.27)右端关于 $n \to 1$ 的极限即为散射数据相应的演化规律.

证明： 设 $\phi(x, k)$ 是线性问题(4.1a)的一个非零解，$\theta(x, k)$ 是与 $\phi(x, k)$ 线性无关的解，则存在常数 γ_1, γ_2 使得下式成立

$$\phi_t(x, k) - N\phi(x, k) = \gamma_1\phi(x, k) + \gamma_2\theta(x, k) \qquad (4.2.28)$$

首先设 k 为离散谱 k_j（$\text{Im} k_j > 0$），在(4.2.28)中选定 $\phi(x, k_j)$ 为相应于 k_j 的归一化特征函数，即：$\phi(x, k_j) = c_j\phi_1(x, k_j)$，$c_j$ 是 $\phi_1(x, k_j)$ 的归一化常数. 当 x 趋于正无穷时，由于 $\phi(x, k_j)$ 以指数衰减，则 $\theta(x, k_j)$ 必以指数增长，知 $\gamma_2 = 0$，于是(4.2.28)化为

$$\phi_t(x, k_j) - N\phi(x, k_j) = \gamma_1\phi(x, k_j) \qquad (4.2.29)$$

以 $(\phi_2(x, k_j), \phi_1(x, k_j))$ 左乘(4.2.29)，有

$$\frac{d}{dt}(\phi_1\phi_2) - (C\phi_1^2 + B\phi_2^2) = 2\gamma_1\phi_1\phi_2 \qquad (4.2.30)$$

上式在整个实轴关于 x 积分并注意到 $2\int_{-\infty}^{\infty} \phi_1\phi_2 dx = 1$，得

$$\gamma_1 = -\int_{-\infty}^{\infty} (C\phi_1^2 + B\phi_2^2)dx \qquad (4.2.31)$$

为方便起见，我们把 γ_1 写成内积的形式

$$\gamma_1 = -<\theta, F> = \int_{-\infty}^{\infty} \theta F \mathrm{d}x, \ \theta = (\phi_2^2, \phi_1^2), \ F = \begin{pmatrix} B \\ C \end{pmatrix}$$

$$(4.2.31')$$

利用引理 4.3、引理 4.4，给出

$$\gamma_1 = -<\theta, F>$$

$$= -\alpha(t) \sum_{j=1}^{m+n-1} (2\mathrm{i}k_j(t))^{m+n-1} <\theta, \widetilde{L}^{j-1}\begin{pmatrix} q \\ r \end{pmatrix}> -$$

$$\beta(t) \sum_{m=1}^{n} (2\mathrm{i}k_j(t))^{n-m} <\theta, \widetilde{L}^{m-1}\begin{pmatrix} xq \\ xr \end{pmatrix}>$$

$$= -\alpha(t) \sum_{j=1}^{m+n-1} (2\mathrm{i}k_j(t))^{m+n-1} <\widetilde{L}^{*\,j-1}\theta, \begin{pmatrix} q \\ r \end{pmatrix}> -$$

$$\beta(t) \sum_{m=1}^{n} (2\mathrm{i}k_j(t))^{n-m} <(\widetilde{L}^*)^{m-1}\theta, \begin{pmatrix} xq \\ xr \end{pmatrix}>$$

$$= -\alpha(t) \sum_{j=1}^{m+n-1} (2\mathrm{i}k_j(t))^{m+n-2} <\theta, \begin{pmatrix} q \\ r \end{pmatrix}> -$$

$$\beta(t) \sum_{m=1}^{n} (2\mathrm{i}k_j(t))^{n-1} <\theta, \begin{pmatrix} xq \\ xr \end{pmatrix}>$$

$$= \frac{1}{2} n\beta(t)(2\mathrm{i}k_j(t))^{n-1} \qquad (4.2.32)$$

所以(4.2.29)成为

$$\phi_t(x, k_j) - N\phi(x, k_j) = \frac{1}{2} n\beta(t)(2\mathrm{i}k_j(t))^{n-1}\phi(x, k_j)$$

$$(4.2.33)$$

注意到当 $x \to +\infty$ 时，

$$N \sim \begin{bmatrix} -\dfrac{1}{2}\alpha(t)(2ik_j(t))^{m+n-1}-\dfrac{1}{2}\beta(t)(2ik_j(t))^n x & 0 \\ 0 & \dfrac{1}{2}\alpha(t)(2ik_j(t))^{m+n-1}+\dfrac{1}{2}\beta(t)(2ik_j(t))^n x \end{bmatrix}$$

$$\phi(x,k_j) \sim c_j(t)\begin{pmatrix}0\\1\end{pmatrix}e^{ik_j(t)x}$$

故得

$$c_{j,t}(t)\begin{pmatrix}0\\1\end{pmatrix}+k_{j,t}(t)\begin{pmatrix}0\\ic_j(t)x\end{pmatrix}-c_j(t)\begin{bmatrix}0\\ \dfrac{1}{2}\alpha(t)(2ik_j(t))^{n+m-1}+\dfrac{1}{2}\beta(t)(2ik_j(t))^n x\end{bmatrix}$$

$$=\frac{1}{2}n\beta(t)(2ik_j(t))^{n-1}c_j(t)\begin{pmatrix}0\\1\end{pmatrix} \tag{4.2.34}$$

即：

$$k_{j,t}(t)=-\frac{1}{2}i\beta(t)(2ik_j(t))^n$$

$$c_{j,t}(t)=\left[\frac{1}{2}\alpha(t)(2ik_j(t))^{m+n-1}+\frac{1}{2}n\beta(t)(2ik_j(t))^{n-1}\right]c_j(t)$$

解此方程即得(4.2.27a, c). 而(4.2.27b, d)可类似得到.

当 k 取实连续谱时,取 $\psi(x,k)$ 为线性问题(4.1a)满足渐进条件(4.1.4a)的解,则(4.2.28)式变为

$$\psi_t(x,k)-N\psi(x,k)=\gamma_1\psi(x,k)+\gamma_2\bar\psi(x,k) \tag{4.2.35}$$

在此等式中令 x 趋于负无穷,利用渐进式(4.1.4a)得：

$$ik_t x\begin{pmatrix}1\\0\end{pmatrix}e^{-ikx}+\left[\frac{1}{2}\alpha(t)(2ik(t))^{m+n-1}+\frac{1}{2}\beta(t)(2ik(t))^n x\right]\begin{pmatrix}1\\0\end{pmatrix}e^{-ikx}$$

$$=\gamma_1\begin{pmatrix}1\\0\end{pmatrix}e^{-ikx}+\gamma_2\begin{pmatrix}0\\-1\end{pmatrix}e^{ikx} \tag{4.2.36}$$

由于函数 $x\mathrm{e}^{\mathrm{i}kx}$, $\mathrm{e}^{\mathrm{i}kx}$ 与 $\mathrm{e}^{-\mathrm{i}kx}$ 线性独立，故有

$$k_t = -\frac{1}{2}\mathrm{i}\beta(t)(2\mathrm{i}k(t))^n, \ \gamma_1 = \frac{1}{2}\alpha(t)(2\mathrm{i}k(t))^{m+n-1}, \ \gamma_2 = 0$$

$$(4.2.37)$$

将 Jost 函数的关系式(4.1.22a)代入(4.2.35)，再令 x 趋于正无穷，
利用渐近式(4.1.2a)给出

$$\begin{pmatrix} \dfrac{\mathrm{d}a(t, k)}{\mathrm{d}t}\mathrm{e}^{-\mathrm{i}kx} \\[2mm] \dfrac{\mathrm{d}b(t, k)}{\mathrm{d}t}\mathrm{e}^{\mathrm{i}kx} \end{pmatrix} - \frac{1}{2}\alpha(t)(2\mathrm{i}k(t))^{m+n-1}\begin{pmatrix} -a(t, k)\mathrm{e}^{-\mathrm{i}kx} \\[2mm] b(t, k)\mathrm{e}^{\mathrm{i}kx} \end{pmatrix}$$

$$= \frac{1}{2}\alpha(t)(2\mathrm{i}k(t))^{m+n-1}\begin{pmatrix} a(t, k)\mathrm{e}^{-\mathrm{i}kx} \\[2mm] b(t, k)\mathrm{e}^{\mathrm{i}kx} \end{pmatrix} \qquad (4.2.38)$$

由此推知

$$\frac{\mathrm{d}a(t, k)}{\mathrm{d}t} = 0, \ \frac{\mathrm{d}b(t, k)}{\mathrm{d}t} = \alpha(t)(2\mathrm{i}k(t))^{m+n-1}b(t, k)$$

$$(4.2.39)$$

从而解得

$$a(t, k) = a(0, k(0)), \ b(t, k) = b(0, k(0))\mathrm{e}^{\int_0^t \alpha(s)(2\mathrm{i}k(s))^{m+n-1}\mathrm{d}s}$$

$$(4.2.40)$$

即可得到(4.2.27e). 而(4.2.27f)可类似得到.

§4.3 AKNS 系统方程族的精确解

本节考虑 AKNS 系统无反射位势 $q(x, t)$ 和 $r(x, t)$，即反射系
数 $R(t, k)$ 和 $\bar{R}(t, k)$ 为零的解.

定理 4.2 设线性问题(4.1a)的散射数据为(4.1.20)，并令

$$F(t, x) = F_c(t, x) + F_d(t, x), \quad \overline{F}(t, x) = \overline{F}_c(t, x) + \overline{F}_d(t, x)$$
$$(4.3.1a, b)$$

其中

$$F_c(t, x) = \frac{1}{2\pi}\int_{-\infty}^{\infty} R(t, k(t))e^{ik(t)x}dk, \quad F_d(t, x) = \sum_{j=1}^{l} c_j^2(t)e^{ik_j(t)x}$$
$$(4.3.2a, b)$$

$$\overline{F}_c(t, x) = \frac{1}{2\pi}\int_{-\infty}^{\infty} \overline{R}(t, k(t))e^{-i\overline{k}(t)x}dk, \quad \overline{F}_d(t, x) = -\sum_{j=1}^{l} \overline{c}_j^2(t)e^{-i\overline{k}_j(t)x}$$
$$(4.3.2c, d)$$

则有

$$q(t, x) = -2K_1(t, x, x), \quad r(t, x) = K_{2,x}(t, x, x)/K_1(t, x, x)$$
$$(4.3.3)$$

其中 $K(t, x, y) = (K_1(t, x, y), K_2(t, x, y))^T$ 满足 GLM 积分方程

$$K(t, x, y) - \binom{1}{0}\overline{F}(t, x+y) + \binom{0}{1}\int_x^{\infty} F(t, z+x)\overline{F}(t, z+y)dz +$$

$$\int_x^{\infty} K(t, x, s)\left(\int_x^{\infty} F(t, z+s)\overline{F}(t, z+y)dz\right)ds = 0 \quad (4.3.4)$$

特别地在无反射情况下，GLM 积分方程简化为

$$K(t, x, y) - \binom{1}{0}\overline{F}_d(t, x+y) + \binom{0}{1}\int_x^{\infty} F_d(t, z+x)\overline{F}_d(t, z+y)dz +$$

$$\int_x^{\infty} K(t, x, s)\left(\int_x^{\infty} F_d(t, z+s)\overline{F}_d(t, z+y)dz\right)ds = 0 \quad (4.3.5)$$

这是退化核的积分方程，写成分量形式

$$K_1(x, y) - \overline{F}_d(x+y) + \int_x^{\infty} K_1(x, s)\left(\int_x^{\infty} F_d(s+z)\overline{F}_d(z+y)dz\right)ds = 0$$

$$(4.3.6a)$$

$$K_2(x, y) + \int_x^\infty F_d(x+z)\overline{F}_d(z+y)\mathrm{d}z +$$

$$\int_x^\infty K_2(x, s)\left(\int_x^\infty F_d(s+z)\overline{F}_d(z+y)\mathrm{d}z\right)\mathrm{d}s = 0 \qquad (4.3.6b)$$

直接计算得

$$\int_x^\infty F_d(s+z)\overline{F}_d(z+y)\mathrm{d}z = -\sum_{j=1}^l \sum_{m=1}^l \frac{\mathrm{i}\, c_j^2(t)\, \overline{c}_m^2(t)}{k_j - \overline{k}_m} \mathrm{e}^{\mathrm{i}k_j(x+s) - \mathrm{i}\overline{k}_m(x+y)}$$

$$(4.3.7)$$

从而可设解向量 $K(t, x, y)$ 具有变量分离的和式

$$K_1(t, x, y) = \sum_{p=1}^{\bar{l}} \overline{c}_p(t) g_p(t, x) \mathrm{e}^{-\mathrm{i}\overline{k}_p y} \qquad (4.3.8a)$$

$$K_2(t, x, y) = \sum_{p=1}^{\bar{l}} \overline{c}_p(t) h_p(t, x) \mathrm{e}^{-\mathrm{i}\overline{k}_p y} \qquad (4.3.8b)$$

将 (4.3.7) 与 (4.3.8) 代入积分方程 (4.3.6) 给出

$$g_m(t, x) + \overline{c}_m(t)\mathrm{e}^{-\mathrm{i}\overline{k}_m x} + \sum_{j=1}^l \sum_{p=1}^{\bar{l}} \frac{c_j^2(t)\, \overline{c}_m(t)\, \overline{c}_p(t)}{(k_j - \overline{k}_m)(k_j - \overline{k}_p)} \mathrm{e}^{\mathrm{i}(2k_j - \overline{k}_m - \overline{k}_p)x} g_p(t, x) = 0$$

$$(4.3.9a)$$

$$h_m(t, x) - \sum_{j=1}^l \frac{\mathrm{i}\, c_j^2(t)\, \overline{c}_m(t)}{k_j - \overline{k}_m} \mathrm{e}^{\mathrm{i}(2k_j - \overline{k}_m)x} +$$

$$\sum_{j=1}^l \sum_{p=1}^{\bar{l}} \frac{c_j^2(t)\, \overline{c}_m(t)\, \overline{c}_p(t)}{(k_j - \overline{k}_m)(k_j - \overline{k}_p)} \mathrm{e}^{\mathrm{i}(2k_j - \overline{k}_m - \overline{k}_p)x} h_p(t, x) = 0,$$

$$(m = 1, 2, \cdots, \bar{l}) \qquad (4.3.9b)$$

若以 $E(t, x)$ 表示 $\bar{l} \times l$ 矩阵, 其元素为

$$e_{mj} = \frac{c_j(t)\, \overline{c}_m(t)}{k_j - \overline{k}_m} \mathrm{e}^{\mathrm{i}(k_j - \overline{k}_m)x} \qquad (4.3.10)$$

则在(4.3.9a)中，未知函数 $g_p(t, x)$ 的系数所构成的矩阵可写成

$$W(t, x) = I + E(t, x)E^T(t, x) \qquad (4.3.11)$$

式中 I 是 $\bar{l} \times \bar{l}$ 单位矩阵，$E^T(t, x)$ 表示 $E(t, x)$ 的转置矩阵. 因此 $W(t, x)$ 是对称矩阵. 再引入两个向量

$$g(t, x) = (g_1(t, x), g_2(t, x), \cdots, g_l(t, x))^T$$
$$\qquad (4.3.12a)$$

$$\bar{\Lambda}(t, x) = (\bar{c}_1(t)e^{-i\bar{k}_1 x}, \bar{c}_2(t)e^{-i\bar{k}_2 x}, \cdots, \bar{c}_l(t)e^{-i\bar{k}_l x})^T$$
$$\qquad (4.3.12b)$$

这时线性方程组(4.3.9a)可简写成矩阵方程

$$W(t, x)g(t, x) = -\bar{\Lambda}(t, x) \qquad (4.3.13)$$

假设 $W(t, x)$ 的逆矩阵存在，由此解得 $g(t, x)$ 并将其代入(4.3.8a)有

$$K_1(t, x, y) = -\bar{\Lambda}^T(t, y)W^{-1}(t, x)\bar{\Lambda}(t, x) \quad (4.3.14)$$

在此等式的两端取迹,应用迹不依赖于矩阵积因子的循环顺序给出

$$K_1(t, x, y) = -tr(W^{-1}(t, x)\bar{\Lambda}(t, x)\bar{\Lambda}^T(t, y)) $$
$$\qquad (4.3.15)$$

对方程(4.3.9b)进行类似的分析解得

$$h(t, x) = iW^{-1}(t, x)E(t, x)\Lambda(t, x) \qquad (4.3.16)$$

其中列向量 $\Lambda(t, x)$ 定义为

$$\Lambda(t, x) = (c_1(t)e^{ik_1 x}, c_2(t)e^{ik_2 x}, \cdots, c_l(t)e^{ik_l x})^T \quad (4.3.17)$$

将(4.3.16)代入(4.3.8b)又给出

$$K_2(t, x, y) = i[tr(W^{-1}(t, x)E(t, x)\Lambda(t, x)\bar{\Lambda}^T(t, y))]$$
$$\qquad (4.3.18)$$

于是按公式(4.2.5)即可确定 AKNS 系统方程族的多孤子解

$$q(t, x) = 2tr(W^{-1}(t, x)\overline{\Lambda}(t, x)\overline{\Lambda}^{\mathrm{T}}(t, x)) \quad (4.3.19a)$$

$$q(t, x)r(t, x) = -2\frac{\mathrm{d}}{\mathrm{d}x}tr\left(W^{-1}(t, x)E(t, x)\frac{\mathrm{d}}{\mathrm{d}x}E^{\mathrm{T}}(t, x)\right)$$

$$(4.3.19b)$$

这里我们给出这些解的一些显示形式.

(1) 当 $l = \bar{l} = 1$，单孤子解为

$$q(t, x) = \frac{2\bar{c}_1^2(t)\mathrm{e}^{-2i\bar{k}_1(t)x}}{1 + \left[\dfrac{c_1(t)\bar{c}_1(t)}{k_1(t) - \bar{k}_1(t)}\right]^2 \mathrm{e}^{2i[k_1(t) - \bar{k}_1(t)]x}} \quad (4.3.20a)$$

$$r(t, x) = \frac{2c_1^2(t)\mathrm{e}^{2ik_1(t)x}}{1 + \left[\dfrac{c_1(t)\bar{c}_1(t)}{k_1(t) - \bar{k}_1(t)}\right]^2 \mathrm{e}^{2i[k_1(t) - \bar{k}_1(t)]x}} \quad (4.3.20b)$$

(2) 在 $l = \bar{l} = 2$ 时，有双孤子解

$$q(t, x) = \frac{\Delta_1}{\det(W(t, x))}, \quad r(t, x) = \frac{\Delta_2}{\det(W(t, x))}$$

$$(4.3.21a, b)$$

其中

$$\Delta_1 = 2\left\{\bar{c}_1^2\mathrm{e}^{-2i\bar{k}_1x} + \bar{c}_2^2\mathrm{e}^{-2i\bar{k}_2x} + \left[\frac{c_1\bar{c}_2\bar{c}_1(\bar{k}_2 - \bar{k}_1)}{(k_1 - \bar{k}_1)(k_1 - \bar{k}_2)}\right]^2\mathrm{e}^{2i(k_1 - \bar{k}_1 - \bar{k}_2)x} + \right.$$

$$\left.\left[\frac{\bar{c}_1c_2\bar{c}_2(\bar{k}_2 - \bar{k}_1)}{(k_2 - \bar{k}_1)(k_2 - \bar{k}_2)}\right]^2\mathrm{e}^{2i(k_2 - \bar{k}_1 - \bar{k}_2)x}\right\}$$

$$\Delta_2 = 2\left\{c_1^2\mathrm{e}^{2ik_1x} + c_2^2\mathrm{e}^{2ik_2x} + \left[\frac{c_1c_2\bar{c}_1(k_2 - k_1)}{(k_1 - \bar{k}_1)(k_2 - \bar{k}_1)}\right]^2\mathrm{e}^{2i(k_1 + k_2 - \bar{k}_1)x} + \right.$$

$$\left[\frac{c_1 c_2 \, \bar{c}_2 (k_2 - k_1)}{(k_1 - \bar{k}_2)(k_2 - \bar{k}_2)}\right]^2 e^{2i(k_1 + k_2 - \bar{k}_2)x}\Bigg\}$$

$$\det(W(t, x)) = 1 + \left[\frac{c_1 \bar{c}_1}{k_1 - \bar{k}_1}\right]^2 e^{2i(k_1 - \bar{k}_1)x} + \left[\frac{c_1 \bar{c}_2}{k_1 - \bar{k}_2}\right]^2 e^{2i(k_1 - \bar{k}_2)x} +$$

$$\left[\frac{c_2 \bar{c}_1}{k_2 - \bar{k}_1}\right]^2 e^{2i(k_2 - \bar{k}_1)x} + \left[\frac{c_2 \bar{c}_2}{k_2 - \bar{k}_2}\right]^2 e^{2i(k_2 - \bar{k}_2)x} +$$

$$\left[\frac{c_1 c_2 \, \bar{c}_1 \, \bar{c}_2 (k_2 - k_1)(\bar{k}_2 - \bar{k}_1)}{(k_1 - \bar{k}_1)(k_1 - \bar{k}_2)(k_2 - \bar{k}_1)(k_2 - \bar{k}_2)}\right]^2 e^{2i(k_1 + k_2 - \bar{k}_1 - \bar{k}_2)x}.$$

c_1，c_2，$\bar{c}_1$，$\bar{c}_2$，k_1，k_2，$\bar{k}_1$ 和 $\bar{k}_2$ 是 t 的函数.

§4.4 约化

本节主要考虑 AKNS 系统方程族的解约化得到等谱 AKNS 方程族、非等谱 AKNS 方程族、AKNS 系统的 τ 方程族和 KdV 系统方程族、mKdV 系统方程族、非线性 Schrödinger 系统方程族、sine-Gordon 系统方程族的解.

§4.4.1 约化为等谱 AKNS 方程族、非等谱 AKNS 方程族以及 τ 方程族的解

（1）当 $\beta(t) = 0$，$\alpha(t) = 1$ 时，散射数据随时间的发展式（4.2.27）变为

$$k_j(t) = k_j(0), \quad \bar{k}_h(t) = \bar{k}_h(0) \qquad (4.4.1\text{a, b})$$

$$c_j(t) = c_j(0) e^{\frac{1}{2}(2ik_j(0))^{m+n-1}t}, \quad \bar{c}_h(t) - \bar{c}_h(0) e^{\frac{1}{2}(2i\bar{k}_h(0))^{m+n-1}t}$$

$$(4.4.1\text{c, d})$$

式中 $k_j(0)$，$\bar{k}_m(0)$，$c_j(0)$ 和 $\bar{c}_m(0)$ 是常数.

将(4.4.1)代入(4.3.19)即得等谱方程族(2.3.12a)的解.

特别当 $l = \bar{l} = 1$ 时，(2.3.12a)的单孤子解为

$$q(t, x) = 2 \frac{\bar{c}_1^2(0) e^{-(2i\bar{k}_1(0))^{m+n-1}t - 2i\bar{k}_1(0)x}}{1 + \frac{c_1^2(0)\,\bar{c}_1^2(0)}{(k_1(0) - \bar{k}_1(0))^2} e^{((2ik_1(0))^{m+n-1} - (2i\bar{k}_1(0))^{m+n-1})t + 2i(k_1(0) - \bar{k}_1(0))x}}$$

$$(4.4.2a)$$

$$r(t, x) = 2 \frac{c_1^2(0) e^{(2ik_1(0))^{m+n-1}t + 2ik_1(0)x}}{1 + \frac{c_1^2(0)\,\bar{c}_1^2(0)}{(k_1(0) - \bar{k}_1(0))^2} e^{((2ik_1(0))^{m+n-1} - (2i\bar{k}_1(0))^{m+n-1})t + 2i(k_1(0) - \bar{k}_1(0))x}}$$

$$(4.4.2b)$$

当取 $m+n-1 = 2, 3$ 时，$c_1(0), \bar{c}_1(0), k_1(0), \bar{k}_1(0)$ 是常数，(4.4.2)就是(2.3.12b, c)的单孤子解,这和[15]结论一致.

(2) 当 $\beta(t) = 1, \alpha(t) = 0$，散射数据随时间的发展式(4.2.27)变为

$$k_j^{1-n}(t) = k_j^{1-n}(0) + (1-n)(2i)^{n-1}t, \quad (j = 1, 2, \cdots, l)$$

$$(4.4.3a)$$

$$\bar{k}_h^{1-n}(t) = \bar{k}_h^{1-n}(0) + (1-n)(2i)^{n-1}t, \quad (h = 1, 2, \cdots, \bar{l})$$

$$(4.4.3b)$$

$$c_j(t) = c_j(0) \cdot [k_j^{1-n}(0) + (1-n)(2i)^{n-1}t]^{\frac{n}{2(1-n)}}, \quad (j = 1, 2, \cdots, l)$$

$$(4.4.3c)$$

$$\bar{c}_h(t) = \bar{c}_h(0)[\bar{k}_h^{1-n}(0) + (1-n)(2i)^{n-1}t]^{\frac{n}{2(1-n)}}, \quad (h = 1, 2, \cdots, \bar{l})$$

$$(4.4.3d)$$

式中 $k_j(0), \bar{k}_m(0), c_j(0)$ 和 $\bar{c}_m(0)$ 是常数.

将(4.4.3)代入(4.3.19),即给出非等谱方程族(2.3.13a)的解.

特别当 $n = 2, l = \bar{l} = 1$ 时非等谱方程方程(2.3.13b)的单孤子解为

$$q(t, x) = 2 \frac{\left(\dfrac{\bar{c}}{\bar{k} - 2\mathrm{i}t}\right)^2 \mathrm{e}^{\frac{-2\mathrm{i}x}{\bar{k} - 2\mathrm{i}t}}}{1 + \left(\dfrac{c\bar{c}}{k - \bar{k}}\right)^2 \mathrm{e}^{\frac{2\mathrm{i}(k - \bar{k})x}{(\bar{k} - 2\mathrm{i}t)(k - 2\mathrm{i}t)}}} \tag{4.4.4a}$$

$$r(t, x) = 2 \frac{\left(\dfrac{c}{k - 2\mathrm{i}t}\right)^2 \mathrm{e}^{\frac{2\mathrm{i}x}{k - 2\mathrm{i}t}}}{1 + \left(\dfrac{c\bar{c}}{k - \bar{k}}\right)^2 \mathrm{e}^{\frac{2\mathrm{i}(\bar{k} - k)x}{(\bar{k} - 2\mathrm{i}t)(k - 2\mathrm{i}t)}}} \tag{4.4.4b}$$

式中 $c, \bar{c}, k$ 和 $\bar{k}$ 是常数;

当 $n = 3, l = \bar{l} = 1$ 时非等谱方程方程(2.3.13c)的单孤子解为

$$q(t, x) = 2 \frac{\dfrac{\bar{c}^2}{(\bar{k} + 8t)^{\frac{3}{2}}} \mathrm{e}^{\frac{-2\mathrm{i}x}{\sqrt{\bar{k} + 8t}}}}{1 + \dfrac{(c\bar{c})^2}{\sqrt{\bar{k} + 8t}\sqrt{k + 8t}(\sqrt{\bar{k} + 8t} - \sqrt{k + 8t})^2} \mathrm{e}^{2\mathrm{i}\frac{\sqrt{k + 8t} - \sqrt{\bar{k} + 8t}}{\sqrt{\bar{k} + 8t}\sqrt{k + 8t}}x}}$$

$$\tag{4.4.5a}$$

$$r(t, x) = 2 \frac{\dfrac{c^2}{(k + 8t)^{\frac{3}{2}}} \mathrm{e}^{\frac{2\mathrm{i}x}{\sqrt{k + 8t}}}}{1 + \dfrac{(c\bar{c})^2}{\sqrt{\bar{k} + 8t}\sqrt{k + 8t}(\sqrt{\bar{k} + 8t} - \sqrt{k + 8t})^2} \mathrm{e}^{2\mathrm{i}\frac{\sqrt{\bar{k} + 8t} - \sqrt{k + 8t}}{\sqrt{\bar{k} + 8t}\sqrt{k + 8t}}x}}$$

$$\tag{4.4.5b}$$

式中 $c, \bar{c}, k$ 和 $\bar{k}$ 是常数.

(3) 约化为 τ 方程族的解

当 $\alpha(t) = mt, \beta(t) = 1$ 时,散射数据的时间发展式(4.2.27)变为

$$k_j(t) = (k_j^{1-n}(0) + (1 - n)(2\mathrm{i})^{n-1} t)^{\frac{1}{1-n}}, \quad (j = 1, 2, \cdots, l) \tag{4.4.6a}$$

$$\bar{k}_h(t) = (\bar{k}_h^{1-n}(0) + (1-n)(2\mathrm{i})^{n-1} t)^{\frac{1}{1-n}}, \ (h = 1, 2, \cdots, \bar{l}) \tag{4.4.6b}$$

$$c_j(t) = \frac{c_j(0)}{k_j^{-\frac{n}{2}}(0)} (k_j^{1-n}(0) + (1-n)(2\mathrm{i})^{n-1}t)^{\frac{n}{2(1-n)}} \mathrm{e}^{\frac{1}{2}T(t)},$$

$$(j = 1, 2, \cdots, l) \tag{4.4.6c}$$

$$\bar{c}_h(t) = \frac{\bar{c}_h(0)}{\bar{k}_h^{-\frac{n}{2}}(0)} (\bar{k}_h^{1-n}(0) + (1-n)(2\mathrm{i})^{n-1}t)^{\frac{n}{2(1-n)}} \mathrm{e}^{\frac{1}{2}\bar{T}(t)},$$

$$(h = 1, 2, \cdots, \bar{l}) \tag{4.4.6d}$$

式中

$$T(t) = (2\mathrm{i})^m t(k_j^{1-n}(0) + (1-n)(2\mathrm{i})^{n-1}t)^{\frac{m}{1-n}} -$$

$$\frac{(2\mathrm{i})^{m+1-n}}{m+1-n} \big[(k_j^{1-n}(0) + (1-n)(2\mathrm{i})^{n-1}t)^{\frac{m+1-n}{1-n}} - k_j^{m+1-n}(0) \big]$$

$$\bar{T}(t) = (2\mathrm{i})^m t(\bar{k}_h^{1-n}(0) + (1-n)(2\mathrm{i})^{n-1}t)^{\frac{m}{1-n}} -$$

$$\frac{(2\mathrm{i})^{m+1-n}}{m+1-n} \big[(\bar{k}_h^{1-n}(0) + (1-n)(2\mathrm{i})^{n-1}t)^{\frac{m+1-n}{1-n}} - \bar{k}_h^{m+1-n}(0) \big]$$

$k_j(0)$, $\bar{k}_h(0)$, $c_j(0)$ 和 $\bar{c}_h(0)$ 是 $(q(0, x), r(0, x))^{\mathrm{T}}$ 时的散射数据.

将(4.4.6)代入(4.3.19),即为 τ 方程族(2.3.14a)的解. 特别当 $n=0$, $m=3$, $l=\bar{l}=1$ 时,第一个非平凡 τ 方程(2.3.14b)的单孤子解为

$$q(t, x) = \frac{2\bar{c}_1 \mathrm{e}^{-(2\mathrm{i}\bar{k}_1+t)x + \frac{1}{4}(2\mathrm{i}\bar{k}_1+t)^3(3t-2\mathrm{i}\bar{k}_1) - \bar{k}_1^4}}{1 + \left(\dfrac{c_1 \bar{c}_1}{k_1 - \bar{k}_1}\right)^2 \mathrm{e}^{2\mathrm{i}(k_1-\bar{k}_1)x + \frac{1}{4}(2\mathrm{i}k_1+t)^3(3t-2\mathrm{i}k_1) + \frac{1}{4}(2\mathrm{i}\bar{k}_1+t)^3(3t-2\mathrm{i}\bar{k}_1) - k_1^4 - \bar{k}_1^4}} \tag{4.4.7a}$$

$$r(t, x) = \frac{2c_1 e^{-(2ik_1+t)x+\frac{1}{4}(2ik_1+t)^3(3t-2ik_1)-k_1^4}}{1+\left(\dfrac{c_1\bar{c}_1}{k_1-\bar{k}_1}\right)^2 e^{2i(k_1-\bar{k}_1)x+\frac{1}{4}(2ik_1+t)^3(3t-2ik_1)+\frac{1}{4}(2i\bar{k}_1+t)^3(3t-2i\bar{k}_1)-k_1^4-\bar{k}_1^4}}$$

$$(4.4.7b)$$

式中 k_1, $\bar{k}_1$, c_1, $\bar{c}_1$ 为常数.

§4.4.2 约化为 mKdV 系统方程族和 KdV 系统方程族的解

一般说来从(2.3.9)不能通过直接约化得到 KdV 系统方程族(2.1.10),因此也就不能直接由 AKNS 系统方程族的解得到 KdV 方程族的解,幸运的是 KdV 系统方程族(2.1.10)和 mKdV 系统方程族(2.2.11)之间存在着著名的 Miura[2,71]变换. 通过 Miura 变换,我们能由 mKdV 系统方程族的解得到 KdV 系统方程族的解.

我们知道当 $(q(t, x), r(t, x))^T = (v(t, x), \pm v(t, x))^T$ 时,奇数阶的 AKNS 系统方程族可约化为 mKdV 系统方程族. 这时. 从(4.1a)和(4.1.23a, b),容易导出下列散射数据的关系:

$$\bar{k}_j(t) = -k_j(t), \bar{c}_j^2(t) = \pm c_j^2(t), \bar{R}(t, k) = \mp R(t, k), \bar{l} = l$$

$$(4.4.8)$$

(一) $(q(t, x), r(t, x))^T = (v(t, x), -v(t, x))^T$ 时,有

$$E(t, x) = \left(\frac{ic_m(t)c_j(t)}{k_m(t)+k_j(t)} e^{i(k_m(t)+k_j(t))x}\right)_{l\times l}$$

$$\bar{\Lambda}(t, x) = i\Lambda(t, x), \bar{\Lambda}(t, x)\bar{\Lambda}^T(t, x) = i\frac{dE(t, x)}{dx}$$

$$(4.4.9)$$

而(4.3.19a)成为

$$v(t, x) = -i\frac{d}{dx}\ln\frac{\det(I-E(t, x))}{\det(I+E(t, x))} \qquad (4.4.10)$$

此即为 mKdV 系统方程族 (2.2.11) 的解(此时 $F = \partial^2 + 4v^2 + 4v_x\partial^{-1}v$,这是需要注意的).

将(4.4.10)代入 Miura 变换

$$u(t,\,x) = \mp iv_x(t,\,x) + v^2(t,\,x) \tag{4.4.11}$$

给出 KdV 系统方程族(2.1.10)的解

$$u(t,\,x) = 2\frac{d^2}{dx^2}\ln\det(I \pm E(t,\,x)) \tag{4.4.12}$$

(1) 在 $\alpha(t) = 1$, $\beta(t) = 0$, $m+n-1 = 3$ 和 $l = 1$ 时, (4.4.10)和(4.4.12)分别成为

$$v(t,\,x) = \frac{-4ic^2 e^{2ikx-8ik^3t}}{1 + \dfrac{c^4}{4k^2}e^{4ikx-16ik^3t}} \tag{4.4.13}$$

和

$$u(t,\,x) = \frac{\mp 4ikc^2 e^{2ikx-8ik^3t}}{1 \pm \dfrac{ic^2}{2k}e^{2ikt-8ik^3t}} \tag{4.4.14}$$

式中 k, c 分别是常数,其分别是等谱 mKdV 方程

$$v_t = v_{xxx} + 6v^2 v_x$$

和等谱 KdV 方程(2.1.13b)的单孤子解

(2) 在 $\alpha(t) = 0$, $\beta(t) = 1$, $m+n-1 = 3$ 和 $l = 1$ 时, (4.4.10)和(4.4.12)分别成为

$$v(t,\,x) = \frac{\dfrac{-2ic^2}{(k+8t)^{\frac{3}{2}}}e^{\frac{2ix}{\sqrt{k+8t}}}}{1 + \dfrac{c^4}{4(k+8t)^2}e^{\frac{4ix}{\sqrt{k+8t}}}} \tag{4.4.15}$$

和

$$u(t,x) = \frac{\mp 16ic^2 e^{\frac{2ix}{\sqrt{k+8t}}}}{(2k+16t \pm ic^2 e^{\frac{2ix}{\sqrt{k+8t}}})^2} \tag{4.4.16}$$

其中 c 和 k 是常数. 它们是非等谱 mKdV 方程

$$v_t = x(v_{xxx} + 6v^2 v_x) + 3v_{xx} + 4v^3 + 2v_x \partial^{-1}v^2$$

和非等谱 KdV 方程(2.1.14b)的解.

(3) 在 $\alpha(t) = 3t$, $\beta(t) = 1$, $m = 3$, $n = 1$, $l = 1$ 时, (4.4.10) 和(4.4.12)分别成为

$$v = \frac{-2ic^2 e^t e^{2ikxe^t + \frac{8}{3}ik^3(e^{3t}-1)-8ik^3 te^{3t}}}{1 + \frac{c^4}{4k^2}e^{4ikxe^t + \frac{16}{3}ik^3(e^{3t}-1)-16ik^3 te^{3t}}} \tag{4.4.17}$$

$$u = \frac{\mp 4ic^2 k e^{2t} e^{2ikxe^t + \frac{8}{3}ik^3(e^{3t}-1)-8ik^3 te^{3t}}}{\left(1 \pm \frac{ic^2}{2k}e^{2ikxe^t + \frac{8}{3}ik^3(e^{3t}-1)-8ik^3 te^{3t}}\right)^2} \tag{4.4.18}$$

它们是 mKdV 系统的 τ 方程

$$v_t = 3t(v_{xxx} + 6v^2 v_x) + (xv)_x$$

和 KdV 系统的 τ 方程(2.1.15b)的解.

(二) $(q(t,x), r(t,x))^T = (v(t,x), v(t,x))^T$ 时, 有

$$E(t,x) = \left(\frac{c_m(t)c_j(t)}{k_m(t)+k_j(t)}e^{i(k_m(t)+k_j(t))x}\right)_{l\times l}$$

$$\bar{\Lambda}(t,x) = \Lambda(t,x), \quad \bar{\Lambda}(t,x)\bar{\Lambda}^T(t,x) = -i\frac{dE(t,x)}{dx} \tag{4.4.19}$$

(4.3.19)成为

$$v(t,\ x) = \frac{\mathrm{d}}{\mathrm{d}x} \ln \frac{\det(I - \mathrm{i}E(t,\ x))}{\det(I + \mathrm{i}E(t,\ x))} \tag{4.4.20}$$

即为 mKdV 系统方程族的解(此时 $F = \partial^2 - 4v^2 - 4v_x \partial^{-1} v$).

将(4.4.20)代入第二个 Miura 变换

$$u(t,\ x) = \mp v_x(t,\ x) - v^2(t,\ x) \tag{4.4.21}$$

得 KdV 系统方程族的解

$$u(t,\ x) = 2 \frac{\mathrm{d}^2}{\mathrm{d}x^2} \ln \det(I \pm \mathrm{i}E(t,\ x)) \tag{4.4.22}$$

(1) 在 $\alpha(t) = 1, \beta(t) = 0, m+n-1 = 3$ 和 $l = 1$ 时,(4.4.20)
和(4.4.22)分别成为

$$v(t,\ x) = \frac{4c^2 \mathrm{e}^{2\mathrm{i}kx - 8\mathrm{i}k^3 t}}{1 + \dfrac{c^4}{4k^2} \mathrm{e}^{4\mathrm{i}kx - 16\mathrm{i}k^3 t}} \tag{4.4.23}$$

和

$$u(t,\ x) = \frac{\mp 4\mathrm{i}kc^2 \mathrm{e}^{2\mathrm{i}kx - 8\mathrm{i}k^3 t}}{1 \pm \dfrac{\mathrm{i}c^2}{2k} \mathrm{e}^{2\mathrm{i}kx - 8\mathrm{i}k^3 t}} \tag{4.4.24}$$

式中 k, c 分别是常数,它们分别是等谱 mKdV 方程(2.2.13b)和等谱
KdV 方程(2.1.13b)的单孤子解

(2) 在 $\alpha(t) = 0, \beta(t) = 1, m+n-1 = 3$ 和 $l = 1$ 时,(4.4.20)
和(4.4.22)分别成为

$$v(t,\ x) = \frac{\dfrac{2c^2}{(k+8t)^{\frac{3}{2}}} \mathrm{e}^{\frac{2\mathrm{i}x}{\sqrt{k+8t}}}}{1 + \dfrac{c^4}{4(k+8t)^2} \mathrm{e}^{\frac{4\mathrm{i}x}{\sqrt{k+8t}}}} \tag{4.4.25}$$

和

$$u(t, x) = \frac{\mp 16\mathrm{i}c^2 \mathrm{e}^{\frac{2\mathrm{i}x}{\sqrt{k+8t}}}}{(2k+16t \pm \mathrm{i}c^2 \mathrm{e}^{\frac{2\mathrm{i}x}{\sqrt{k+8t}}})^2} \tag{4.4.26}$$

其中 c 和 k 是常数. 其分别是下面非等谱 mKdV 方程(2.2.14b)和非等谱 KdV 方程(2.1.14b)的解.

(3) 在 $\alpha(t) = 3t$, $\beta(t) = 1$, $m = 3$, $n = 1$, $l = 1$ 时, (4.4.20)和(4.4.22)分别成为

$$v = \frac{2c^2 \mathrm{e}^t \mathrm{e}^{2\mathrm{i}kx\,\mathrm{e}^t + \frac{8}{3}\mathrm{i}k^3(\mathrm{e}^{3t}-1) - 8\mathrm{i}k^3 t\mathrm{e}^{3t}}}{1 + \frac{c^4}{4k^2}\mathrm{e}^{4\mathrm{i}kx\,\mathrm{e}^t + \frac{16}{3}\mathrm{i}k^3(\mathrm{e}^{3t}-1) - 16\mathrm{i}k^3 t\mathrm{e}^{3t}}} \tag{4.4.27}$$

$$u = \frac{\mp 4\mathrm{i}c^2 k\mathrm{e}^{2t} \mathrm{e}^{2\mathrm{i}kx\,\mathrm{e}^t + \frac{8}{3}\mathrm{i}k^3(\mathrm{e}^{3t}-1) - 8\mathrm{i}k^3 t\mathrm{e}^{3t}}}{\left(1 \pm \frac{\mathrm{i}c^2}{2k}\mathrm{e}^{2\mathrm{i}kx\,\mathrm{e}^t + \frac{8}{3}\mathrm{i}k^3(\mathrm{e}^{3t}-1) - 8\mathrm{i}k^3 t\mathrm{e}^{3t}}\right)^2} \tag{4.4.28}$$

其为 mKdV 系统 τ 方程(2.2.15b)和 KdV 系统的 τ 方程(2.1.15b)的解.

§4.4.3　约化为非线性 Schrödinger 系统方程族的解

在第二章§2.3.2.2,在谱参数 k 满足 $k_t = \frac{\beta(t)}{2}(2k)^2$ 和边值条件(2.3.18), $r = \mp q^*$ 时, AKNS 方程族可化为非线性 Schrödinger 系统方程族,此时定理 4.1 中离散散射数据随时间演化规律变为

$$k_j(t) = \left[k_j^{1-n} + (1-n)2^{n-1}\int_0^t \beta(s)\mathrm{d}s\right]^{\frac{1}{1-n}} \tag{4.4.29a}$$

$$c_j(t) = c_j(0)\mathrm{e}^{\frac{1}{2}\left[\int_0^t (\mathrm{i}\alpha(s)(2k_j(s))^{m+n-1} + n\beta(s)(2k_j(s))^{n-1})\mathrm{d}s\right]} \tag{4.4.29b}$$

下面考虑非线性 Schrödinger 方程族的解

$(q(t, x), r(t, x))^{\mathrm{T}} = (v(t, x), -v^*(t, x))^{\mathrm{T}}$, 从(4.1a)和

(4.1.23a，b)给出

$$\bar{k}_j = k_j^*,\ \bar{c}_j^2 = -c_j^{*2},\ \bar{R}(k) = R^*(k),\ \bar{l} = l,\ \bar{\Lambda}(t, x) = -\mathrm{i}\Lambda^*(t, x)$$

$$E(t, x) = \left[\frac{\mathrm{i}c_j(t)c_m^*(t)}{k_j(t) - k_m^*(t)}\mathrm{e}^{\mathrm{i}(k_j(t) - k_m^*(t))x}\right]_{l\times l}$$

$$W(t, x) = I + E(t, x)E^*(t, x) \qquad (4.4.30)$$

则，(4.3.19)成为

$$v(t, x) = -2tr\left(W^{-1}(t, x)\Lambda^*(t, x)\Lambda^{*T}(t, x)\right) \qquad (4.4.31a)$$

$$v(t, x)v^*(t, x) = 2\frac{\mathrm{d}}{\mathrm{d}x}tr\left(W^{-1}(t, x)E(t, x)\frac{\mathrm{d}}{\mathrm{d}x}E^*(t, x)\right) \qquad (4.4.31b)$$

即为非线性 Schrödinger 方程族的解.

(1) 在 $\alpha(t) = 1$，$\beta(t) = 0$，$m+n-1 = 2$ 和 $l = 1$ 时，(4.4.31a)为

$$v(t, x) = \frac{-2c^{*2}\mathrm{e}^{-2\mathrm{i}k^*x - 4\mathrm{i}k^{*2}t}}{1 - \left(\frac{cc^*}{k - k^*}\right)^2\mathrm{e}^{8\mathrm{i}(k^2 - k^{*2})t + 2\mathrm{i}(k - k^*)x}} \qquad (4.4.32)$$

其中 c 和 k 是常数. 这恰好是非线性 Schrödinger 方程 $v_t = \mathrm{i}v_{xx} + 2\mathrm{i}v|v|^2$ 的解.

(2) 在 $\alpha(t) = 0$，$\beta(t) = 1$，$n = 2$ 和 $l = 1$ 时，(4.4.31a)成为

$$v(t, x) = \frac{-2\dfrac{c^{*2}}{(k^* - 2t)^2}\mathrm{e}^{\frac{-2\mathrm{i}x}{k^* - 2t}}}{1 - \left(\dfrac{cc^*}{k - k^*}\right)^2\dfrac{\mathrm{e}^{\frac{2\mathrm{i}(k^* - k)x}{(k - 2t)(k^* - 2t)}}}{}} \qquad (4.4.33)$$

其中 c 和 k 是常数. 为非等谱非线性 Schrödinger 方程

$$v_t = \mathrm{i}xv_{xx} + 2\mathrm{i}v_x - 2\mathrm{i}xv\mid v\mid^2 - 2\mathrm{i}v\partial^{-1}\mid v\mid^2$$

的解.

(3) 在 $\alpha(t) = mt$, $\beta(t) = 1$, $m = 2$, $n = 1$ 和 $l = 1$ 时, (4.4.31a)
成为

$$v(t,\ x) = \frac{-2c^{*2}\mathrm{e}^{t+4\mathrm{i}k^{*2}te^{2t}-2\mathrm{i}k^{*2}(e^{2t}-1)-2\mathrm{i}k^*xe^t}}{1-\left(\dfrac{cc^*}{k-k^*}\right)^2\mathrm{e}^{2\mathrm{i}(k-k^*)xe^t+2\mathrm{i}(k^2-k^{*2})(e^{2t}-1)-4\mathrm{i}(k^2-k^{*2})te^{2t}}}$$

$$(4.4.34)$$

其中 c 和 k 是常数,为非线性 Schöodinger 系统的 τ 方程

$$v_t = -2\mathrm{i}t(v_{xx}+2v\mid v\mid^2)+(xv)_x$$

的解.

§4.4.4 约化为 sine-Gordon 系统方程族的解

最后,考虑如何约化为 sine-Gordon 系统方程族的解.

根据对约化为 mKdV 系统方程族的分析,sine-Gordon 方程族和
mKdV 方程族之间存在紧密联系,通过取 $(q(t,\ x),\ r(t,\ x))^{\mathrm{T}} = \frac{1}{2}(u_x(t,\ x),\ -u_x(t,\ x))^{\mathrm{T}}$ 和以 $-n$, $-m$ 代替 n, m,则得 sine-Gordon 系统方程族(2.3.22). 由 §4.4.2(一),有

$$E(t,\ x) = \left(\frac{\mathrm{i}c_m(t)c_j(t)}{k_m(t)+k_j(t)}\mathrm{e}^{\mathrm{i}(k_m(t)+k_j(t))x}\right)_{l\times l}$$

$$\bar{\Lambda}(t,\ x) = \mathrm{i}\Lambda(t,\ x),\ \bar{\Lambda}(t,\ x)\bar{\Lambda}^{\mathrm{T}}(t,\ x) = \mathrm{i}\frac{\mathrm{d}E(t,\ x)}{\mathrm{d}x}$$

方程族(2.3.22)的解为

$$u(t,\ x) = -2\mathrm{i}\ln\frac{\det(I-E(t,\ x))}{\det(I+E(t,\ x))} \qquad (4.4.35)$$

（1）在 $\alpha(t)=1$，$\beta(t)=0$，$n=1$，$l=1$ 时，(4.4.35)成为 sine-Gordon 方程(2.2.27b)的单孤子解

$$u=-2\mathrm{i}\ln\frac{1-\dfrac{\mathrm{i}c^2}{2k}\mathrm{e}^{2\mathrm{i}kx-\frac{\mathrm{i}t}{2k}}}{1+\dfrac{\mathrm{i}c^2}{2k}\mathrm{e}^{2\mathrm{i}kx-\frac{\mathrm{i}t}{2k}}} \tag{4.4.36}$$

其中 c 和 k 是常数.

（2）在 $\alpha(t)=0$，$\beta(t)=1$ $n=1$ 和 $l=1$ 时，(4.4.35)成为非等谱 sine-Gordon 方程(2.2.28b)的单孤子解

$$u(t,x)=2\mathrm{i}\ln\frac{1+\dfrac{\mathrm{i}c^2}{2k-t}\mathrm{e}^{\mathrm{i}x\sqrt{4k-2t}}}{1-\dfrac{\mathrm{i}c^2}{2k-t}\mathrm{e}^{\mathrm{i}x\sqrt{4k-2t}}}=-4\arctan\mathrm{e}^{\omega+\mathrm{i}x\sqrt{4k-2t}} \tag{4.4.37}$$

其中 c 和 k 是常数，$\mathrm{e}^{\omega}=\dfrac{c^2}{2k-t}$.

（3）在 $\alpha(t)=t$，$\beta(t)=1$，$m=1$，$n=-1$，$l=1$ 时，(4.4.35)成为 τ 方程(2.2.29b)的解

$$u(t,x)=2\mathrm{i}\ln\frac{1+\dfrac{\mathrm{i}c^2}{2\sqrt{k-\frac{1}{2}t}}\mathrm{e}^{2\mathrm{i}x\sqrt{k-\frac{1}{2}t}+2\mathrm{i}t\sqrt{k-\frac{1}{2}t}+\frac{8\mathrm{i}}{3}\left(k-\frac{1}{2}t\right)^{\frac{3}{2}}-\frac{1}{2}\ln(4k-2t)}}{1-\dfrac{\mathrm{i}c^2}{2\sqrt{k-\frac{1}{2}t}}\mathrm{e}^{2\mathrm{i}x\sqrt{k-\frac{1}{2}t}+2\mathrm{i}t\sqrt{k-\frac{1}{2}t}+\frac{8\mathrm{i}}{3}\left(k-\frac{1}{2}t\right)^{\frac{3}{2}}-\frac{1}{2}\ln(4k-2t)}} \tag{4.4.38}$$

其中 c 和 k 是常数.

第五章　一些非线性发展方程的双线性形式和 Wronskian 形式解

双线性导数方法是求解一大类非线性发展方程的强有力工具，可系统地构造 N - 孤子解、Bäcklund 变换解、无穷守恒律等[81]；Wronskian 技巧由于行列式特有的性质可以方便地将解代入方程进行验证. 在本章中，首先介绍双线性导数和 Wronski 行列式的定义与性质，然后通过双线性导数方法和 Wronskian 技巧获得非等谱 sine-Gordon 方程、非等谱非线性 Schrödinger 方程的解. 同时用双线性导数方法获得 KdV 系统的 τ 方程、非线性 Schrödinger 系统的 τ 方程以及 sine-Gordon 系统的 τ 方程的解.

§5.1　双线性导数和 Wronski 行列式

§5.1.1　双线性导数的定义及性质

定义 5.1　设 $f(t, x)$ 与 $g(t, x)$ 是变量 t 与 x 的可微函数，m 和 n 是任意的非负整数，定义双线性导数为

$$D_t^m D_x^n f \cdot g = (\partial_t - \partial_{t'})^m (\partial_x - \partial_{x'})^n f(t, x) g(t', x') \big|_{t'=t, \, x'=x}$$

$$(5.1.1)$$

(1) 由定义可推出一些常用双线性导数的性质

$$D_x^m a \cdot 1 = \partial^m a / \partial x^m \qquad (5.1.2a)$$

$$D_x^m a \cdot b = (-1)^m D_x^m b \cdot a \qquad (5.1.2b)$$

$$D_x^m a \cdot a = 0, \, m \text{ 为奇数} \qquad (5.1.2c)$$

$$D_x D_t a \cdot 1 = D_x D_t 1 \cdot a = \frac{\partial^2 a}{\partial x \partial t} \quad (5.1.2d)$$

$$D_x^m a \cdot b = 0 \longleftrightarrow a = \infty b \quad (5.1.2e)$$

$$D_x^{2m+1} a_t \cdot a = \frac{1}{2} D_x^{2m+1} D_t a \cdot a \quad (5.1.2f)$$

（2）双线性导数的一些基本性质

$$\exp(D_x) a(x) \cdot b(x) = a(x + \epsilon) b(x - \epsilon) \quad (5.1.3a)$$

$$D_x^m \exp(p_1 x) \cdot \exp(p_2 x) = (p_1 - p_2)^m \exp(p_1 + p_2)x$$
$$(5.1.3b)$$

$$\exp(\epsilon D_x + \delta D_t) a \cdot b = \exp[\sinh(\epsilon \partial/\partial x + \delta \partial/\partial t) \ln(a/b) +$$
$$\cosh(\epsilon \partial/\partial x + \delta \partial/\partial t) \ln(ab)]$$
$$(5.1.3c)$$

§5.1.2　Wronski 行列式的定义与性质

定义 5.2　一组可微函数 $(\phi_1, \phi_2, \cdots, \phi_N)^{\mathrm{T}}$ 的 N 阶 Wronski 行列式定义为

$$W(\phi_1, \phi_2, \cdots, \phi_N) = \begin{vmatrix} \phi_1 & \phi_1^{(1)} & \cdots & \phi_1^{(N-1)} \\ \vdots & \vdots & & \vdots \\ \phi_N & \phi_N^{(1)} & \cdots & \phi_N^{(N-1)} \end{vmatrix} \quad (5.1.4)$$

其中 $\phi_j^{(l)} = \partial^l \phi_j / \partial x^l$. 一般将其写成紧凑格式

$$W = |\phi, \phi^{(1)}, \cdots, \phi^{(N-1)}| = |0, 1, \cdots, N-1| = |\widehat{N-1}| \quad (5.1.5)$$

下列是几种常见的紧凑记法

1) $|-1, \widehat{N-3}, N-1, N+1| = |\phi^{(-1)}, \phi, \cdots, \phi^{(N-3)}, \phi^{(N-1)}|,$

$\phi^{(N+1)} \mid$

其中 $\phi_j^{(-1)} = \partial^{-1}\phi_j$，$\partial^{-1}$ 为积分算子，$\partial^{-1}\partial = \partial\partial^{-1} = 1$.

2) $\mid \hat{l}_1, l_2, \cdots, l_p \mid = \mid \phi, \phi^{(1)}, \cdots, \phi^{(l_1)}, \phi^{(l_2)}, \cdots, \phi^{(l_p)} \mid$,

3) $\mid \tilde{h}_1, h_2, \cdots, h_q \mid = \mid \phi^{(1)}, \cdots, \phi^{(h_1)}, \phi^{(h_2)}, \cdots, \phi^{(h_q)} \mid$.

Wronski 行列式 (5.1.4) 后一列是前一列导数的特点，使 Wronski 行列式按列求导变得非常便利. 例如：若 $f = \mid \widehat{N-1} \mid$，则 $f_x = \mid \widehat{N-2}, N \mid$，$f_{xx} = \mid \widehat{N-3}, N-1, N \mid + \mid \widehat{N-2}, N+1 \mid$.

将线性代数中行列式的一些恒等式应用到 Wronski 行列式 (5.1.4)，可得 Wronskian 技巧中一些常用恒等式.

(1) 设矩阵 $A = (a_{ij})_{N\times N} = [\alpha_1, \alpha_2, \cdots, \alpha_N]$. $\alpha_j = (a_{1j}, a_{2j}, \cdots, a_{nj})^T$ 为 A 的列向量，向量 $b = (b_1, b_2, \cdots, b_N)^T$，则成立

$$\sum_{j=1}^N \mid \alpha_1, \cdots, \alpha_{j-1}, b\alpha_j, \alpha_{j+1}, \cdots, \alpha_N \mid = \sum_{j=1}^N b_j \mid A \mid$$

(5.1.6)

其中，$b\alpha_j$ 表示 $(b_1 a_{1j}, b_2 a_{2j}, \cdots, b_N a_{Nj})^T$.

(2) 假设 $A = (a_{ij})_{N\times N}$ 是 $N\times N$ 矩阵，α_j 表示其列向量，β_j 表示行向量. $P = (P_{ij})_{N\times N}$ 是 $N\times N$ 算子矩阵，即：P_{ij} 是算子，则

$$\sum_{j=1}^N \mid \alpha_1, \cdots, \alpha_{j-1}, C_j\alpha_j, \alpha_{j+1}, \cdots, \alpha_N \mid = \sum_{j=1}^N \begin{vmatrix} \beta_1 \\ \vdots \\ \beta_{s-1} \\ R_s\beta_s \\ \beta_{s+1} \\ \vdots \\ \beta_N \end{vmatrix}$$

(5.1.7)

其中 $C_j\alpha_j = (P_{1j}a_{1j}, \cdots, P_{Nj}a_{Nj})$，$R_s\beta_s = (P_{s1}\alpha_{s1}, \cdots, P_{sN}\alpha_{sN})$.

这与[82]和[83]相比,更为一般,可通过行列式展开性质得到证明.

(3) 设 Wronski 行列式(5.1.4)中 ϕ_j 满足

$$\phi_{j,\,xx} = k_j^2 \phi_j \qquad (5.1.8\text{a})$$

有

$$(\sum_{j=1}^{N} k_j^2)|\,\widehat{N-1}\,| = -|\,\widehat{N-3,\,N-1,\,N}\,| + |\,\widehat{N-2,\,N+1}\,|$$

$$(5.1.8\text{b})$$

$$(\sum_{j=1}^{N} k_j^2)|\,\widehat{N-2,\,N}\,| = -|\,\widehat{N-4,\,N-2,\,N-1,\,N}\,| + |\,\widehat{N-2,\,N+2}\,|$$

$$(5.1.8\text{c})$$

$$(\sum_{j=1}^{N} k_j^2)|\,\widetilde{N}\,| = -|\,\widetilde{N-2,\,N,\,N+1}\,| + |\,\widetilde{N-1,\,N+2}\,|$$

$$(5.1.8\text{d})$$

(4) 若 Wronski 行列式(5.1.6)满足条件(5.1.8a),则基于等式

$$|\,\widehat{N-1}\,|\{(\sum_{j=1}^{N} k_j^2)[(\sum_{j=1}^{N} k_j^2)|\,\widehat{N-1}\,|]\} = [(\sum_{j=1}^{N} k_j^2)|\,\widehat{N-1}\,|]^2$$

$$(5.1.9\text{a})$$

可得

$$|\,\widehat{N-1}\,|(|\,\widehat{N-5,\,N-3,\,N-2,\,N-1,\,N}\,| - |\,\widehat{N-4,\,N-2,\,N-1}$$

$$N+1\,| + 2\,|\,\widehat{N-3,\,N,\,N+1}\,| - |\,\widehat{N-3,\,N-1,\,N+2}\,| + |\,\widehat{N-2}$$

$$N+3\,|) = (-|\,\widehat{N-3,\,N-1,\,N}\,| + |\,\widehat{N-2,\,N+1}\,|)^2$$

$$(5.1.9\text{b})$$

(5) 若记 M 为 $N \times (N-2)$ 矩阵,a, b, c 和 d 都是 N 维列向量,则成立

$$|M,\,a,\,b\,||M,\,c,\,d\,| - |M,\,a,\,c\,||M,\,b,\,d\,| + |M,\,a,\,d\,||M,\,b,\,c\,| = 0$$

$$(5.1.10)$$

§5.1.3　双 Wronski 行列式的定义

定义 5.3　称 $(M+N)\times(M+N)$ 阶 Wronski 行列式，

$$|\,\widehat{M-1}\,;\,\widehat{N-1}\,|=|\,\phi,\,\partial_x^1\phi,\,\cdots,\,\partial_x^{M-1}\phi\,;\,\psi,\,\partial_x^1\psi,\,\cdots,\,\partial_x^{N-1}\psi\,|$$

$$(5.1.11)$$

其中 ϕ 和 ψ 是如下 $(M+N)$ 列向量

$$\phi=(\phi_1,\,\phi_2,\,\cdots,\,\phi_{M+N})^{\mathrm{T}},\ \psi=(\psi_1,\,\psi_2,\,\cdots,\,\psi_{M+N})^{\mathrm{T}}$$

称这种特殊的 Wronski 行列式为双 Wronski 行列式.

如果 $M=0$，$(5.1.11)$ 就是一般的 $N\times N$ 阶 Wronski 行列式；如果 $N=0$，$(5.1.11)$ 就是一般的 $M\times M$ 阶 Wronski 行列式.

§5.2　非等谱 sine-Gordon 方程的解

§5.2.1　双线性导数形式的解

对于非等谱 sine-Gordon 方程[109](2.2.28b)

$$v_{xt}=(\cos v\partial^{-1}\cos v\partial^{-1}+\sin v\partial^{-1}\sin v\partial^{-1})(xv_x)_x \quad (5.2.1)$$

引入变换

$$v=2\mathrm{i}\ln\frac{f^*}{f} \quad (5.2.2)$$

方程 $(5.2.1)$ 可化为

$$\mathrm{i}\frac{1}{f^{*2}}D_xD_tf^*\cdot f^*-\mathrm{i}\frac{1}{f^2}D_xD_tf\cdot f=\frac{x}{2\mathrm{i}}\Big(\frac{f^2}{f^{*2}}-\frac{f^{*2}}{f^2}\Big)+$$

$$\frac{1}{2\mathrm{i}}\frac{f^2}{f^{*2}}\partial^{-1}\frac{f^{*2}-f^2}{f^2}-\frac{1}{2\mathrm{i}}\frac{f^{*2}}{f^2}\partial^{-1}\frac{f^2-f^{*2}}{f^{*2}} \quad (5.2.3)$$

引入辅助函数 $g(t,\,x)$，使得

$$\frac{f^2 - f^{*2}}{f^{*2}} = \left(\frac{g}{f^*}\right)_x \qquad (5.2.4)$$

则,得(5.2.1)的双线性导数方程

$$D_x D_t f \cdot f - \frac{x}{2}(f^2 - f^{*2}) + \frac{1}{2}f^* g = 0 \qquad (5.2.5a)$$

$$D_x g \cdot f^* + f^{*2} - f^2 = 0 \qquad (5.2.5b)$$

将 f 和 g 做摄动展开

$$f = 1 + f^{(1)}\varepsilon + f^{(2)}\varepsilon^2 + f^{(3)}\varepsilon^3 + \cdots \qquad (5.2.6a)$$

$$g = g^{(1)}\varepsilon + g^{(2)}\varepsilon^2 + g^{(3)}\varepsilon^3 + \cdots \qquad (5.2.6b)$$

代入(5.2.5),比较 ε 的同次幂系数给出

$$2f_{xt}^{(1)} - x(f^{(1)} - f^{*(1)}) + \frac{1}{2}g^{(1)} = 0 \qquad (5.2.7a)$$

$$2f_{xt}^{(2)} - x(f^{(2)} - f^{*(2)}) + \frac{1}{2}g^{(2)} = \frac{x}{2}(f^{(1)}f^{(1)} - f^{*(1)}f^{*(1)}) -$$

$$\frac{1}{2}f^{*(1)}g^{(1)} - D_x D_t f^{(1)} \cdot f^{(1)} \qquad (5.2.7b)$$

$$\cdots\cdots$$

$$g_x^{(1)} - 2(f^{*(1)} - f^{(1)}) = 0 \qquad (5.2.8a)$$

$$g_x^{(2)} - 2f^{(2)} + 2f^{*(2)} = -D_x g^{(1)} \cdot f^{*(1)} + f^{(1)}f^{(1)} - f^{*(1)}f^{*(1)}$$

$$(5.2.8b)$$

$$\cdots\cdots$$

令

$$f^{(1)} = \omega_1(t)e^{k_1(t)x + \xi_1^{(0)} + \frac{i\pi}{2}} \qquad (5.2.9)$$

代入(5.2.8a)得

$$g^{(1)} = 4 \frac{\omega_1(t)}{k_1(t)} e^{k_1(t)x + \xi_1^{(0)} + \frac{i\pi}{2}} \tag{5.2.10}$$

将(5.2.9)、(5.2.10)代入(5.2.7a)，给出

$$2\omega_1' k_1 + 2x\omega_1 k_1 k_1' + 2\omega_1 k_1' - 2x\omega_1 + 2\frac{\omega_1}{k_1} = 0 \tag{5.2.11}$$

因此有

$$k_1' = \frac{1}{k_1}, \qquad \omega_1' = -2\frac{\omega_1}{k_1^2} \tag{5.2.12}$$

解此方程，得

$$k_1(t) = \sqrt{2t + c_1}, \qquad \omega_1(t) = \frac{1}{2t + c_1}, c_1 \text{ 是常数}$$

$$\tag{5.2.13}$$

将 $f^{(1)}$，$g^{(1)}$ 代入(5.2.7b)，(5.2.8b)可依次推出

$$f^{(j)} = 0, \ g^{(j)} = 0, \ j = 2, 3, \cdots \tag{5.2.14}$$

即 f，g 的摄动展开被截断为有限项. 令 $\varepsilon = 1$，得

$$f = 1 + f^{(1)} = 1 + \omega_1(t) e^{k_1(t)x + \xi_1^{(0)} + \frac{i\pi}{2}}, \ g = 4 \frac{\omega_1(t)}{k_1(t)} e^{k_1(t)x + \xi_1^{(0)} + \frac{i\pi}{2}}$$

$$\tag{5.2.15}$$

所以(5.2.1)的单孤子解为

$$v = 2i\ln \frac{1 + \omega_1(t) e^{k_1(t)x - \frac{i\pi}{2}}}{1 + \omega_1(t) e^{k_1(t)x + \frac{i\pi}{2}}} = 4\arctan e^{\xi_1 + \ln \omega_1(t)} \tag{5.2.16}$$

式中 $\xi_1(t) = k_1(t)x + \xi_1^{(0)}$.

　　若令

$$f^{(1)} = \omega_1(t) e^{k_1(t)x + \xi_1^{(0)} + \frac{i\pi}{2}} + \omega_2(t) e^{k_2(t)x + \xi_2^{(0)} + \frac{i\pi}{2}} \quad (5.2.17)$$

代入(5.2.8a),有

$$g^{(1)} = 4 \frac{\omega_1(t)}{k_1(t)} e^{k_1(t)x + \xi_1^{(0)} + \frac{i\pi}{2}} + 4 \frac{\omega_2(t)}{k_2(t)} e^{k_2(t)x + \xi_2^{(0)} + \frac{i\pi}{2}} \quad (5.2.18)$$

将(5.2.17),(5.2.18)代入(5.2.8a,b)可得

$$f^{(2)} = \left(\frac{k_1(t) - k_2(t)}{k_1(t) + k_2(t)} \right)^2 \omega_1 \omega_2 e^{k_1(t)x + \xi_1^{(0)} + k_2(t)x + \xi_2^{(0)} + i\pi}$$

$$(5.2.19a)$$

$$g^{(3)} = 4 \left(\frac{k_1(t) - k_2(t)}{k_1(t) + k_2(t)} \right)^2 \frac{k_1 + k_2}{k_1 k_2} \omega_1 \omega_2 e^{k_1(t)x + \xi_1^{(0)} + k_2(t)x + \xi_2^{(0)}}$$

$$(5.2.19b)$$

同样由 $f^{(1)}$, $f^{(2)}$, $g^{(1)}$, $g^{(3)}$ 出发,可依次推出

$$f^{(j)} = 0, \ g^{(j)} = 0, \ j = 3, 4, \cdots \quad (5.2.20)$$

即 f, g 的摄动展开被截断为有限项. 令 $\varepsilon = 1$,得(5.2.1)的双孤子解为

$$v = 2i \ln \frac{1 + \omega_1(t) e^{\xi_1 - \frac{i\pi}{2}} + \omega_2(t) e^{\xi_2 - \frac{i\pi}{2}} + \left(\frac{k_1(t) - k_2(t)}{k_1(t) + k_2(t)} \right)^2 \omega_1 \omega_2 e^{\xi_1 + \xi_2 - i\pi}}{1 + \omega_1(t) e^{\xi_1 + \frac{i\pi}{2}} + \omega_2(t) e^{\xi_2 + \frac{i\pi}{2}} + \left(\frac{k_1(t) - k_2(t)}{k_1(t) + k_2(t)} \right)^2 \omega_1 \omega_2 e^{\xi_1 + \xi_2 + i\pi}}$$

$$= 4 \arctan \frac{\omega_1(t) e^{\xi_1} + \omega_2(t) e^{\xi_2}}{1 - \omega_1(t) \omega_2(t) e^{\xi_1 + \xi_2 + A_{12}}} \quad (5.2.21)$$

式中 $\xi_j = k_j(t)x + \xi_j^{(0)}$, $(j = 1, 2)$, $e^{A_{12}} = \left[\frac{k_1(t) - k_2(t)}{k_1(t) + k_2(t)} \right]^2$.

若取

$$f^{(1)} = \sum_{j=1}^{N} \omega_j(t) e^{\xi_j + \frac{\pi}{2}i}, \quad \xi_j = k_j(t)x + \xi_j^{(0)} \tag{5.2.22}$$

式中 $\xi_j^{(0)}$ 是实常数. 类似计算, 得

$$f = \sum_{\varepsilon=0,1} \exp\left[\sum_{j=1}^{N} \varepsilon_j \left(\xi_j + \frac{\pi}{2}i + \ln \omega_j(t) \right) + \sum_{1 \leqslant j < l}^{N} \varepsilon_j \varepsilon_l A_{jl} \right] \tag{5.2.23a}$$

$$k_j(t) = \sqrt{c_j + 2t}, \quad \omega_j(t) = \frac{1}{c_j + 2t}, \quad e^{A_{jl}} = \left[\frac{k_l(t) - k_j(t)}{k_l(t) + k_j(t)} \right]^2 \tag{5.2.23b}$$

这里 c_j 是实参数. 非等谱 sine-Gordon 方程(5.2.1)的 N 孤子解为

$$v = 2i\ln \frac{\sum_{\varepsilon=0,1} \exp\left[\sum_{j=1}^{N} \varepsilon_j \left(\xi_j - \frac{\pi}{2}i + \ln \omega_j(t) \right) + \sum_{1 \leqslant j < l}^{N} \varepsilon_j \varepsilon_l A_{jl} \right]}{\sum_{\varepsilon=0,1} \exp\left[\sum_{j=1}^{N} \varepsilon_j \left(\xi_j + \frac{\pi}{2}i + \ln \omega_j(t) \right) + \sum_{1 \leqslant j < l}^{N} \varepsilon_j \varepsilon_l A_{jl} \right]} \tag{5.2.24}$$

§5.2.2 Wronskian 形式的解

定理5.1 非等谱 sine-Gordon 方程的双线性(5.2.5)具有 Wronskian 解

$$f = |0, 1, 2, \cdots, N-1| = |\widehat{N-1}| \tag{5.2.25a}$$

$$g = -\prod_{j=1}^{N} \frac{1}{k_j(t)} |0, 2, 3, \cdots, N| \tag{5.2.25b}$$

这里 ϕ_j 满足条件

$$\phi_{j,x} = k_j(t)\phi_j^* \tag{5.2.26a}$$

$$\phi_{j,t} = \frac{x}{4}\partial^{-1}\phi_j \tag{5.2.26b}$$

证明:对(5.2.26a)取共轭并对 x 积分一次

$$\phi_j^* = k_j(t)\partial^{-1}\phi_{j,x}$$

(5.2.26b)两端对 x 求 l 阶导数,得

$$\frac{\partial^{l+1}\phi_j}{\partial x^l \partial t} = \frac{l}{4}\frac{\partial^{l-2}\phi_j}{\partial x^{l-2}} + \frac{x}{4}\frac{\partial^{l-1}\phi_j}{\partial x^{l-1}}, \; (l \geqslant 1), \frac{\partial^{-1}}{\partial x^{-1}} = \partial^{-1}$$

则有

$$f_x = |\widehat{N-2}, N|$$

$$f_t = \frac{x}{4}|-1, \widetilde{N-1}| - \frac{1}{4}|-1, 0, 2, \cdots, N-1|$$

$$f_{tx} = \frac{x}{4}(|\widehat{N-1}| + |-1, \widetilde{N-2}, N|) - \frac{1}{4}|-1, 0, 2, \cdots, N-2, N|$$

和

$$f^* = \prod_{j=1}^{N} k_j |-1, \widehat{N-2}| = \prod_{j=1}^{N} \frac{1}{k_j} |\widetilde{N}|, \; f^{*2} = |-1, \widetilde{N-2}| \cdot |\widetilde{N}|$$

$$\bar{g} = -|-1, \widetilde{N-1}|$$

那么,类似于[84]的步骤,将其代入(5.2.5)的左端,利用 Wronskian 行列式的性质(5.1.10)容易证明 f 和 g 是(5.2.5)的解.

满足(5.2.26)的 ϕ_j 为

$$\phi_j = e^{\frac{\pi i}{4}}[e^{\frac{\xi_j}{2} + \frac{\pi i}{4}} \pm e^{\frac{\xi_j}{2} - \frac{\pi i}{4}}], \; (j = 1, 2, \cdots, N) \quad (5.2.27)$$

式中 $\xi_j = k_j(t)x + \xi_j^0$,$k_j(t)$ 由(5.2.23b)给出.

Wronskian 形式的一孤子解和二孤子解为:

$$v = 4\arctan[e^{(x\sqrt{c_1 + 2t} + \xi_j^0)}] \quad (5.2.28a)$$

$$v = 4\arctan \frac{(k_1(t) + k_2(t))(\mathrm{e}^{\frac{1}{2}(\xi_1 - \xi_2)} + \mathrm{e}^{\frac{1}{2}(\xi_2 - \xi_1)})}{(k_1(t) - k_2(t))(\mathrm{e}^{\frac{1}{2}(\xi_1 + \xi_2)} - \mathrm{e}^{\frac{1}{2}(-\xi_1 - \xi_2)})}$$

$$(5.2.28b)$$

§5.2.3 解的性质

一、Hirota 解的性质

以一孤子解和二孤子解为例来考察非等谱方程解的特征. 当 $N = 1$, (5.2.23a)为

$$f = 1 + \omega_1(t)\mathrm{e}^{\xi_1 + \frac{\pi}{2}\mathrm{i}} \qquad (5.2.29)$$

由(5.2.2)知非等谱 sine-Gordon 方程(5.2.1)的单孤子解为

$$v = 2\mathrm{i}\ln \frac{1 + \omega_1(t)\mathrm{e}^{\xi_1 - \frac{\pi}{2}\mathrm{i}}}{1 + \omega_1(t)\mathrm{e}^{\xi_1 + \frac{\pi}{2}\mathrm{i}}} = 4\arctan \mathrm{e}^{\Lambda_1}, \quad \Lambda_1 = \xi_1 + \ln \omega_1(t)$$

$$(5.2.30)$$

为比较等谱与非等谱解的差别,给出 sine-Gordon 方程 $u_{xt} = \sin u$ 的单孤子解[85]i. e. ,

$$u = 2\ln \frac{1 - \mathrm{i}\mathrm{e}^{\theta_1}}{1 + \mathrm{i}\mathrm{e}^{\theta_1}} = 4\arctan \mathrm{e}^{\theta_1}, \quad \theta_1 = h_1 x - \frac{1}{h_1}t + \theta_1^{(0)}$$

$$(5.2.31)$$

其中 h_1 和 $\theta_1^{(0)}$ 是实常数.

图 6 给出了等谱与非等谱波的形状和运动情况.

图 6(a)是 sine-Gordon 方程的一孤子解(5.2.31),该孤波具有

(1) 不变的振幅 2π,因为 e^{θ_1} 总是大于 0;

(2) 给定时刻 t,波的拐点满足

$$\frac{\partial^2}{\partial x^2}u = 4\,h_1^2\,\mathrm{e}^{\theta_1}\,\frac{1-\mathrm{e}^{2\theta_1}}{(1+\mathrm{e}^{2\theta_1})^2} = 0 \qquad (5.2.32)$$

即,

$$\theta_1 = h_1\,x - \frac{1}{h_1}t + \theta_1^{(0)} = 0 \qquad (5.2.33)$$

拐点在平面 $u = \pi$ 上沿特征直线(5.2.33)以定常速度 $1/h_1^2$ 传播;

(3)波在拐点处的斜率

$$\left.\frac{\partial u}{\partial x}\right|_{\theta_1=0} = 2\,h_1 \qquad (5.2.34)$$

不随时间改变,即波形在传播中保持不变.

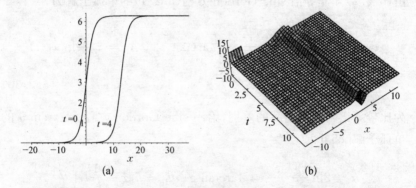

(a)　　　　　　　　　　(b)

图 6　sine-Gordon 方程的一孤子解和非等谱 sine-Gordon 方程的类孤子解

(a) $h_1 = 0.5$，$\theta_1^{(0)} = 0$，一孤子解(5.2.31)的形状和运动;

(b) $c_1 = 0.5$，$\xi_1^{(0)} = 0$，一类孤子解(5.2.30)的形状和运动.

图 6(b)是非等谱 sine-Gordon 方程的 1-类孤子解,该波的特点

(1)波幅和等谱情形一样仍是 2π;

(2)给定 t,波的拐点在平面 $v = \pi$ 上,沿曲线

$$\Lambda_1 = 0 \qquad (5.2.35)$$

运动. 运动速度为

$$\frac{\mathrm{d}x(t)}{\mathrm{d}t} = \frac{2 - \ln(c_1 + 2t)}{(c_1 + 2t)^{\frac{3}{2}}} \qquad (5.2.36)$$

当 $-c_1/2 < t < (e^2 - c_1)/2$，波速为正，向右传播；当 $t > (e^2 - c_1)/2$，波速变为负值，改变传播方向；当 $t \to +\infty$，波速趋向于零，即波达到稳定态.

（3）波在拐点处的斜率

$$\frac{\partial v}{\partial x}\bigg|_{\Lambda_1 = 0} = 2\sqrt{c_1 + 2t} \qquad (5.2.37)$$

随时间变化，即波形随时间变化，当 $t \to +\infty$，拐点处的斜率趋向于 $+\infty$. 这些和等谱既相似又有区别的特征，是我们称其为类孤子解的原因.

图 7(a) 给出了 sine-Gordon 方程 2-孤子解[85]：

$$u = 4\arctan \frac{e^{h_1 x + h_1^{-1} t} + e^{h_2 x + h_2^{-1} t}}{1 - \left(\dfrac{h_1 - h_2}{h_1 + h_2}\right)^2 e^{(h_1 + h_2)x + (h_1^{-1} + h_2^{-1})t}} \qquad (5.2.38)$$

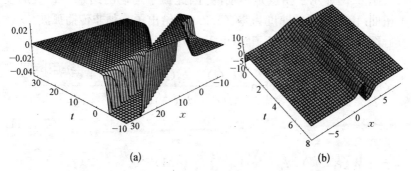

(a)   (b)

图 7 等谱方程的 2-孤子解和非等谱方程的 2-类孤子解

(a) $h_1 = 0.5$, $h_2 = 1$ 等谱方程 2-孤子解的形状；

(b) $c_1 = 0.5$, $c_2 = 1$, $\xi_1^{(0)} = \xi_2^{(0)} = 0$ 非等谱方程 2-类孤子解的形状.

的相互作用情况. 图 7(b)给出了非等谱 sine-Gordon 方程 2-类
孤子解

$$v = 4\arctan\frac{\omega_1(t)e^{\xi_1} + \omega_2(t)e^{\xi_2}}{1 - \omega_1(t)\omega_2(t)e^{\xi_1+\xi_2+A_{12}}} \tag{5.2.39}$$

的相互作用.

二、Wronskian 解的性质

对于等谱情形, Hirota 形式的解和 Wronskian 形式的解是一致
的[86, 87]. 然而, 对于非等谱情形, 两种形式的解具有本质的差别.
这是非等谱方程所特有的性质. 其 Wronskian 形式的解可以得到非
传播孤子波, 非传播孤子波不仅实验室被成功地观察到[88, 89]. 而
且在理论上也得到了证实[90].

首先考察 Wronskian 形式的 1-类孤子解

$$u = 2\ln\frac{e^{-\frac{\pi i}{4}}\left[e^{\frac{\xi_1}{2}-\frac{\pi i}{4}} + e^{-\frac{\xi_1}{2}+\frac{\pi i}{4}}\right]}{e^{\frac{\pi i}{4}}\left[e^{\frac{\xi_1}{2}+\frac{\pi i}{4}} + e^{-\frac{\xi_1}{2}-\frac{\pi i}{4}}\right]} = 4\arctan e^{\xi_1} \tag{5.2.40}$$

图 8(a)给出了 1-类孤子解的形状和运动.

图 8(b)显示了拐点是不动的, 因此孤子运动的结果导致波形发
生扭曲. 由于拐点的速度为零, (5.2.40)给出了一种非传播孤波.

图 9 给出了 2-类孤子解(5.2.41)的相互作用

$$v = 2\mathrm{i}\ln\frac{\bar f}{f} = 4\arctan\frac{(k_1(t)+k_2(t))(e^{\frac{1}{2}(\xi_1-\xi_2)} + e^{\frac{1}{2}(\xi_1-\xi_2)})}{(k_1(t)-k_2(t))(e^{\frac{1}{2}(\xi_1+\xi_2)} - e^{\frac{1}{2}(-\xi_1-\xi_2)})} \tag{5.2.41}$$

$$f = W(\phi_1, \phi_2), \phi_j = e^{\frac{\pi i}{4}}\left[e^{\frac{\xi_j}{2}+\frac{\pi i}{4}} + (-1)^{j+1}e^{-\frac{\xi_j}{2}-\frac{\pi i}{4}}\right], (j = 1, 2)$$

$\xi_j, k_j(t)$ $(j = 1, 2)$ 由(5.2.22)和(5.2.23b)给出.

另一个有趣的性质是初相位决定波的传播, 这是与等谱情形完

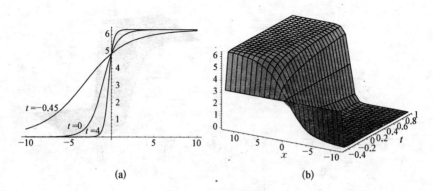

(a)   (b)

图 8 $c_1 = 1$, $\xi_1^{(0)} = 0$, 1-类孤子解(5.2.40)的形状和运动

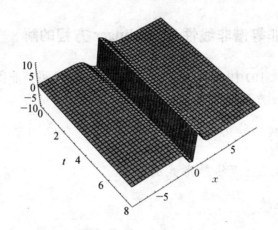

图 9 $c_1 = 6$, $c_2 = 4$, $\xi_1^{(0)} = \xi_2^{(0)} = 0$, 2-类孤子解
(5.2.41)的形状和运动

全不同的特征. 如果我们取 $\xi_1^{(0)} = 1$, 拐点的速度不再为零, 因为

$$\frac{\mathrm{d}x(t)}{\mathrm{d}t} = (c_1 + 2t)^{-\frac{3}{2}}$$

然而仍有一个不动点, 不过, 这个不动点不再是拐点. 见图 10.

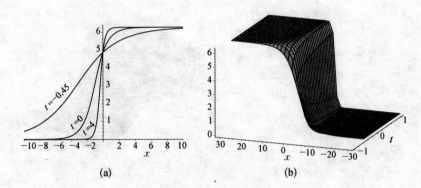

图 10 $c_1 = 1, \xi_1^{(0)} = 1$, 1-类孤子解 $(5.2.40)$ 的波形及运动

§5.3　非等谱非线性 Schrödinger 方程的解

在 $(2.3.19)$ 中，令 $\alpha(t) = 0$，$\beta(t) = 1$，$n = 2$ 得到非等谱非线性 Schrödinger 方程

$$v_t = \mathrm{i}xv_{xx} + 2\mathrm{i}v_x - 2\mathrm{i}xv \mid v \mid^2 - 2\mathrm{i}v\partial^{-1} \mid v \mid^2 \quad (5.3.1)$$

§5.3.1　双线性形式的解

引入变换

$$v = \frac{g}{f}, f \text{ 是实函数} \quad (5.3.2)$$

方程 $(5.3.1)$ 可化为

$$\frac{1}{f^2}D_t g \cdot f = \mathrm{i}x \frac{1}{f^2}\left(D_x^2 g \cdot f + 2\frac{gf_x^2 - gff_{xx} + g^2 g^*}{f}\right) + 2\mathrm{i}\frac{1}{f^2}g_x f$$
$$\qquad\qquad\qquad\qquad\qquad (5.3.3)$$

令

$$D_x^2 f \cdot f = 2gg^* \quad (5.3.4)$$

得方程(5.3.1)的双线性方程

$$(iD_t + xD_x^2)g \cdot f + 2g_x f = 0 \tag{5.3.5a}$$

$$D_x^2 f \cdot f = 2gg^* \tag{5.3.5b}$$

将 f 和 g 作摄动展开

$$f = 1 + \varepsilon^2 f^{(2)} + \varepsilon^4 f^{(4)} + \cdots \tag{5.3.6a}$$

$$g = \varepsilon g^{(1)} + \varepsilon^3 g^{(3)} + \cdots \tag{5.3.6b}$$

代入双线性方程(5.3.5),比较 ε 的同次幂系数,有

$$ig_t^{(1)} + xg_{xx}^{(1)} + 2g_x^{(1)} = 0 \tag{5.3.7a}$$

$$ig_t^{(3)} + xg_{xx}^{(3)} + 2g_x^{(3)} = -(iD_t + xD_x^2)g^{(1)} \cdot f^{(2)} - 2g_x^{(1)} f^{(2)} \tag{5.3.7b}$$

$$ig_t^{(5)} + xg_{xx}^{(5)} + 2g_x^{(5)} = -(iD_t + xD_x^2)(g^{(3)} \cdot f^{(2)} + g^{(1)} \cdot f^{(4)}) -$$
$$2g_x^{(3)} f^{(2)} - 2g_x^{(1)} f^{(4)} \tag{5.3.7c}$$

$$\cdots$$

$$2f_{xx}^{(2)} = 2g^{(1)} g^{(1)*} \tag{5.3.8a}$$

$$2f_{xx}^{(4)} = -D_x^2 f^{(2)} \cdot f^{(2)} + 2g^{(1)} g^{(3)*} + 2g^{(3)} g^{(1)*} \tag{5.3.8b}$$

$$2f_{xx}^{(6)} = -D_x^2 f^{(2)} \cdot f^{(4)} + 2g^{(1)} g^{(5)*} + 2g^{(3)} g^{(3)*} + 2g^{(5)} g^{(1)*} \tag{5.3.8c}$$

$$\cdots$$

设

$$g^{(1)} = e^{\xi_1}, \quad \xi_1 = k_1(t)x + \omega_1(t) + \xi_1^{(0)} \tag{5.3.9}$$

代入(5.3.7a)给出

$$i(k_1'(t)x + \omega_1'(t)) + xk_1^2(t) + 2k_1(t) = 0 \qquad (5.3.10)$$

因此

$$k_1'(t) = ik_1^2(t), \ \omega_1'(t) = 2ik_1(t) \qquad (5.3.11a, b)$$

解此方程,得

$$k_1(t) = \frac{1}{k_1(0) - it}, \ \omega_1(t) = -2\ln(k_1(0) - it)$$

$$(5.3.12a, b)$$

将(5.3.9)代入(5.3.8a),有

$$f^{(2)} = \frac{1}{(k_1(t) + k_1^*(t))^2} e^{\xi_1 + \xi_1^*} \qquad (5.3.13)$$

把 $g^{(1)}$, $f^{(2)}$ 分别代入(5.3.7b,c)和(5.3.8b,c),可依次推知

$$g^{(j)} = 0, \ f^{(j+1)} = 0, \ (j = 3, 5, \cdots) \qquad (5.3.14)$$

即 f, g 的摄动展开可以截断为有限项.

令 $\varepsilon = 1$,得

$$f = 1 + \frac{1}{(k_1(t) + k_1^*(t))^2} e^{\xi_1 + \xi_1^*}, \ g = e^{\xi_1} \quad (5.3.15a,b)$$

非等谱非线性 Schrödinger 方程的单孤子解为

$$v = \frac{g}{f} = \frac{e^{\xi_1}}{1 + \dfrac{1}{(k_1(t) + k_1^*(t))^2} e^{\xi_1 + \xi_1^*}} \qquad (5.3.16)$$

若取 $g^{(1)} = e^{\xi_1} + e^{\xi_2}$, $\xi_j = k_j(t)x + \omega_j(t) + \xi_j^{(0)}$, $j = 1, 2$

$$(5.3.17)$$

代入(5.3.8a)有

$$f^{(2)} = \frac{e^{\xi_1 + \xi_1^*(t)}}{(k_1(t) + k_1^*(t))^2} + \frac{e^{\xi_1 + \xi_2^*}}{(k_1(t) + k_2^*(t))^2} + \frac{e^{\xi_2 + \xi_1^*}}{(k_2(t) + k_1^*(t))^2} +$$

$$\frac{e^{\xi_2 + \xi_2^*}}{(k_2(t) + k_2^*(t))^2} \tag{5.3.18}$$

将(5.3.17)，(5.3.18)代入(5.3.7b)可给出

$$g^{(3)} = \frac{(k_1(t) - k_2(t))^2 e^{\xi_1 + \xi_1^* + \xi_2}}{(k_2(t) + k_1^*(t))^2 (k_1(t) + k_1^*(t))^2} +$$

$$\frac{(k_1(t) - k_2(t))^2 e^{\xi_1 + \xi_2 + \xi_2^*}}{(k_2(t) + k_2^*(t))^2 (k_1(t) + k_2^*(t))^2} \tag{5.3.19}$$

又由(5.3.8b)，得

$$f^{(4)} = \frac{(k_2(t) - k_1(t))^2 (k_2^*(t) - k_1^*(t))^2}{(k_1 + k_1^*)(k_1 + k_2^*)(k_2 + k_1^*)(k_2 + k_2^*)} e^{\xi_1 + \xi_2 + \xi_1^* + \xi_2^*}$$

$$\tag{5.3.20}$$

继续代入后续方程，可给出 $g^{(2j+1)} = 0$，$f^{(2j+2)} = 0$，$j = 2, 3, \cdots$.

即 f, g 的摄动展开可以截断为有限项

$$f = 1 + \frac{e^{\xi_1 + \xi_1^*}}{(k_1(t) + k_1^*(t))^2} + \frac{e^{\xi_1 + \xi_2^*}}{(k_1(t) + k_2^*(t))^2} +$$

$$\frac{e^{\xi_2 + \xi_1^*}}{(k_2(t) + k_1^*(t))^2} + \frac{e^{\xi_2 + \xi_2^*}}{(k_2(t) + k_2^*(t))^2} +$$

$$\frac{(k_2(t) - k_1(t))^2 (k_2^*(t) - k_1^*(t))^2}{(k_1(t) + k_1^*(t))(k_1(t) + k_2^*(t))(k_2(t) + k_1^*(t))(k_2(t) + k_2^*(t))} e^{\xi_1 + \xi_2 + \xi_1^* + \xi_2^*}$$

$$\tag{5.3.21a}$$

$$g = e^{\xi_1} + e^{\xi_2} + \frac{(k_1(t) - k_2(t))^2 e^{\xi_1 + \xi_1^* + \xi_2}}{(k_2(t) + k_1^*(t))^2 (k_1(t) + k_1^*(t))^2} +$$

$$\frac{(k_1(t) - k_2(t))^2 e^{\xi_1 + \xi_2 + \xi_2^*}}{(k_2(t) + k_2^*(t))^2 (k_1(t) + k_2^*(t))^2} \tag{5.3.21b}$$

非等谱非线性 Schrödinger 方程的双孤子解为由 $v = \dfrac{g}{f}$ 给出.

若取

$$g^{(1)} = e^{\xi_1 + \xi_2 + \cdots + \xi_n}, \ \xi_j = k_j(t)x + \omega_j(t) + \xi_j^{(0)}, \ \xi_j^{(0)} 是常数 \tag{5.3.22}$$

可得

$$f = \sum_{\mu = 0, 1} A_1(\mu) e^{\sum\limits_{j=1}^{2n} \mu_j \xi_j + \sum\limits_{1 \leqslant j < l}^{2n} \mu_j \mu_l \theta_{jl}} \tag{5.3.23a}$$

$$g = \sum_{\mu = 0, 1} A_2(\mu) e^{\sum\limits_{j=1}^{2n} \mu_j \xi_j + \sum\limits_{1 \leqslant j < l}^{2n} \mu_j \mu_l \theta_{jl}} \tag{5.3.23b}$$

其中

$$k_j(t) = \frac{1}{k_j(0) - \mathrm{i}t}, \ \omega_j = -2\ln(k_j(0) - \mathrm{i}t), \ (j = 1, 2, \cdots, n) \tag{5.3.24a}$$

$$\xi_{n+j} = \xi_j^*, \ (j = 1, 2, \cdots, n) \tag{5.3.24b}$$

$$e^{\theta_{j(n+l)}} = \frac{1}{(k_j(t) + k_l^*(t))}, \ (j, l = 1, 2, \cdots, n) \tag{5.3.24c}$$

$$e^{\theta_{jl}} = (k_j(t) - k_l(t))^2, \ (j < l = 2, 3, \cdots, n) \tag{5.3.24d}$$

$$e^{\theta_{(n+j)(n+l)}} = (k_j^*(t) - k_l^*(t))^2 \tag{5.3.24e}$$

而 $A_1(\mu)$ 与 $A_2(\mu)$ 表示当 $\mu_j(j = 1, 2, \cdots, n)$ 取所有可能的 0 或 1 时，还需要分别满足条件

$$\sum_{j=1}^{n} \mu_j = \sum_{j=1}^{n} \mu_{n+j}, \ \sum_{j=1}^{n} \mu_j = \sum_{j=1}^{n} \mu_{n+j} + 1 \qquad (5.3.25)$$

由(5.3.2)知方程(5.3.1)的 N 孤子解为

$$v = \frac{\sum_{\mu=0,1} A_2(\mu) e^{\sum_{j=1}^{2n} \mu_j \xi_j + \sum_{1 \leqslant j < l}^{2n} \mu_j \mu_l \theta_{jl}}}{\sum_{\mu=0,1} A_1(\mu) e^{\sum_{j=1}^{2n} \mu_j \xi_j + \sum_{1 \leqslant j < l}^{2n} \mu_j \mu_l \theta_{jl}}} \qquad (5.3.26)$$

§5.3.2 双 Wronskian 形式的解

定理 5.2 非等谱非线性 Schrödinger 方程的双线性方程(5.3.5)具有 $2N \times 2N$ 阶双 Wronskain 解

$$f = |\ \widehat{N-1};\ \widehat{N-1}\ | \qquad (5.3.27a)$$

$$g = 2\ |\ \widehat{N}\ \widehat{N-2}\ | \qquad (5.3.27b)$$

其中

$$\phi_{j,x} = -\frac{\mathrm{i} k_j(t)}{2} \phi_j, \ \psi_{j,x} = \frac{\mathrm{i} k_j(t)}{2} \psi_j, \ j = 1, 2, \cdots, N$$
$$(5.3.28a)$$

$$\phi_{j,x} = \frac{\mathrm{i} k_j^*(t)}{2} \phi_j, \ \psi_{j,x} = \frac{-\mathrm{i} k_j^*(t)}{2} \psi_j, \ j = N+1, N+2, \cdots, 2N$$
$$(5.3.28b)$$

$$\phi_{j,t} = 2\mathrm{i} x \phi_{j,xx} - 2\mathrm{i}(N-1)\phi_{j,x}, \ j = 1, 2, \cdots, 2N$$
$$(5.3.28c)$$

$$\psi_{j,t} = -2\mathrm{i} x \psi_{j,xx} + 2\mathrm{i}(N-1)\psi_{j,x}, \ j = 1, 2, \cdots, 2N$$
$$(5.3.28d)$$

$$k_j(t) = \frac{1}{k_j(0) - \mathrm{i} t}, \ k_j(0) \text{ 是复常数} \qquad (5.3.28e)$$

证明：对 $(5.3.28c,d)$ 两端求 x 的 l 阶导数，得

$$\phi_{j,t}^{(l)} = 2\mathrm{i}x\phi^{(l+2)} - 2\mathrm{i}(N-1-l)\phi^{(l+1)}, \ j = 1, 2, \cdots, 2N$$

$$(5.3.29a)$$

$$\psi_{j,t}^{(l)} = -2\mathrm{i}x\psi^{(l+2)} + 2\mathrm{i}(N-1-l)\psi^{(l+1)}, \ j = 1, 2, \cdots, 2N$$

$$(5.3.29b)$$

利用行列式的性质，由 $(5.3.29)$ 不难算得

$$f_t = 2\mathrm{i}x(|\ \widehat{N-2}, N+1;\ \widehat{N-1}\ | + |\ \widehat{N-1};\ \widehat{N-3}, N-1, N\ | -$$

$$|\ \widehat{N-3}, N-1, N;\ \widehat{N-1}\ | - |\ \widehat{N-1};\ \widehat{N-2}, N+1\ |)$$

$$(5.3.30a)$$

$$g_t = 4\mathrm{i}x(|\ \widehat{N-1}, N+2;\ \widehat{N-2}\ | + |\ \widehat{N-2}, N+1, N;\ \widehat{N-2}\ | - \hat{N};$$

$$\widehat{N-3}, N\ | - |\ \hat{N};\ \widehat{N-4}, N-1, N-2\ |) + 4\mathrm{i}\ |\ \widehat{N-1},$$

$$N+1;\ \widehat{N-2}\ | + 4\mathrm{i}\ |\ \hat{N};\ \widehat{N-3}, N-1\ | \qquad (5.3.30b)$$

$$f_x = |\ \widehat{N-2}, N;\ \widehat{N-1}\ | + |\ \widehat{N-1};\ \widehat{N-2}, N\ |$$

$$(5.3.31a)$$

$$f_{xx} = |\ \widehat{N-3}, N-1, N;\ \widehat{N-1}\ | + |\ \widehat{N-2}, N+1;\ \widehat{N-1}\ | +$$

$$2\ |\ \widehat{N-2}, N;\ \widehat{N-2}, N\ | + |\ \widehat{N-1};\ \widehat{N-3}, N-1,$$

$$N\ | + |\ \widehat{N-1};\ \widehat{N-2}, N+1\ | \qquad (5.3.31b)$$

$$g_x = 2\ |\ \widehat{N-1}, N+1;\ \widehat{N-2}\ | + 2\ |\ \hat{N};\ \widehat{N-3}, N-1\ |$$

$$(5.3.31c)$$

$$g_{xx} = 2\ |\ \widehat{N-2}, N, N+1;\ \widehat{N-2}\ | + 2\ |\ \widehat{N-1}, N+2;\ \widehat{N-2}\ | +$$

$$2\ |\ \hat{N};\ \widehat{N-3} + 2\ |\ \hat{N};\ \widehat{N-4}, N-2, N-1\ |, N\ | + 4\ |\ \widehat{N-1},$$

$$N+1; \widehat{N-3}, N-1 \mid \tag{5.3.31d}$$

则

$$
\begin{aligned}
f_{xx}f - f_x^2 - gg^* = &\mid \widehat{N-1}; \widehat{N-1} \mid (\mid \widehat{N-3}, N-1, N; \widehat{N-1} \mid + \\
&\mid \widehat{N-2}, N+1; \widehat{N-1} \mid + 2 \mid \widehat{N-2}, \\
&N; \widehat{N-2}, N \mid + \mid \widehat{N-1}; \widehat{N-3}, N-1, N \mid + \\
&\mid \widehat{N-1}; \widehat{N-2}, N+1 \mid) - (\mid \widehat{N-2}, N; \\
&\widehat{N-1} \mid + \mid \widehat{N-1}; \widehat{N-2}, N \mid)^2 - 4 \mid \hat{N}; \\
&\widehat{N-2} \mid \mid \widehat{N-2}; \hat{N} \mid = 4(\mid \widehat{N-2}, N; \widehat{N-2}, \\
&N \mid \mid \widehat{N-1}; \widehat{N-1} \mid - \mid \widehat{N-2}, N; \widehat{N-1} \mid \mid \\
&\widehat{N-1}; \widehat{N-2}, N \mid - \mid \hat{N}; \widehat{N-2} \mid \mid \widehat{N-2}; \hat{N} \mid)
\end{aligned}
\tag{5.3.32}
$$

根据行列式的性质 5 知(5.3.32)为零,即双 Wronskian 形式的 f 和 g 满足双线性方程(5.3.5b).

为验证双 Wronskian 形式的 f 和 g 满足双线性方程(5.3.5a),我们给出几个关系式

$$\left(-\sum_{j=1}^{N}\left(\frac{k_j^2}{4} + \frac{k_j^{*2}}{4}\right)f\right)g = f\left(-\sum_{j=1}^{N}\left(\frac{k_j^2}{4} + \frac{k_j^{*2}}{4}\right)g\right) \tag{5.3.33a}$$

$$
\begin{aligned}
\left(\sum_{j=1}^{N}\left(-\frac{\mathrm{i}k_j}{2} + \frac{\mathrm{i}k_j^*}{2}\right) \mid \widehat{N-1}, N+1; \widehat{N-2} \mid\right)f = &\mid \widehat{N-1}, \\
N+1; \widehat{N-2} \mid \sum_{j=1}^{N}&\left(-\frac{\mathrm{i}k_j}{2} + \frac{\mathrm{i}k_j^*}{2}\right)f
\end{aligned}
\tag{5.3.33b}
$$

$$\left(\sum_{j=1}^{N}\left(-\frac{\mathrm{i}k_j}{2} + \frac{\mathrm{i}k_j^*}{2}\right) \mid \widehat{N-2}, N; \widehat{N-1} \mid\right)g = \mid \widehat{N-2},$$

$$N;\ \widehat{N-1}\ |\ \Big(\sum_{j=1}^{N}\Big(-\frac{\mathrm{i}k_j}{2}+\frac{\mathrm{i}k_j^*}{2}\Big)g \qquad (5.3.33\mathrm{c})$$

即

$$2(|\ \widehat{N-2},\ N+1;\ \widehat{N-1}\ |+|\ \widehat{N-3},\ N,\ N-1;\ \widehat{N-1}\ |+|\ \widehat{N-1};$$

$$\widehat{N-3},\ N,\ N-1\ |+|\ \widehat{N-1};\ \widehat{N-2},\ N+1\ |)\ |\ \hat{N};\ \widehat{N-2}\ |$$

$$=2\ |\ \widehat{N-1};\ \widehat{N-1}\ |\ (|\ \widehat{N-1},\ N+2;\ \widehat{N-2}\ |-|\ \widehat{N-2},\ N,$$

$$N+1;\ \widehat{N-2}\ |+|\ \hat{N};\ \widehat{N-3},\ N\ |-|\ \hat{N};\ \widehat{N-4},\ N-2,\ N-1\ |)$$

$$\qquad\qquad (5.3.34\mathrm{a})$$

$$(|\ \widehat{N-2},\ N,\ N+1;\ \widehat{N-2}\ |+|\ \widehat{N-1},\ N+2;\ \widehat{N-2}\ |-|\ \widehat{N-1},$$

$$N+1;\ \widehat{N-3},\ N-1\ |)\ |\ \widehat{N-1};\ \widehat{N-1}\ |$$

$$=|\ \widehat{N-1},\ N+1;\ \widehat{N-2}\ |\ (|\ \widehat{N-2},\ N;\ \widehat{N-1}\ |-|\ \widehat{N-1};$$

$$\widehat{N-2},\ N\ |)\cdot \qquad\qquad (5.3.34\mathrm{b})$$

$$2(|\ \widehat{N-3},\ N-1,\ N;\ \widehat{N-1}\ |+|\ \widehat{N-2},\ N+1;\ \widehat{N-1}-|\ \widehat{N-2},$$

$$N;\ \widehat{N-2},\ N\ |)\ |\ \hat{N};\ \widehat{N-2}\ |=|\ \widehat{N-2},\ N;\ \widehat{N-1}\ |\ (|\ \widehat{N-1},$$

$$N+1;\ \widehat{N-2}\ |-|\ \hat{N};\ \widehat{N-3},\ N-1\ |) \qquad (5.3.34\mathrm{c})$$

下面验证 f、g 满足(5.3.5a)

将(5.3.30)，(5.3.31)代入(5.3.5a)容易算得不显含 x 的项系数为零，因此只需验证显含 x 部分为零，即验证

$$\mathrm{i}(g_t f-g f_t)+x(g_{xx}f-2g_x f_x+g f_{xx})$$

中显含 x 的部分

$$2f(3\ |\ \widehat{N-2},\ N,\ N+1;\ \widehat{N-2}\ |+3\ |\ \hat{N};\ \widehat{N-3},\ N\ |+2\ |\ \widehat{N-1},$$

$$N+1;\ \widehat{N-3},\ N-1\ |-|\ \widehat{N-1},\ N+2;\ \widehat{N-2}\ |-|\ \hat{N};\ \widehat{N-4},$$

$$N-2, N-1|)+2(3|\widehat{N-2}, N+1; \widehat{N-1}|+3|\widehat{N-1}; \widehat{N-3},$$

$$N-1, N|-|\widehat{N-1}; \widehat{N-2}, N+1|-|\widehat{N-3}, N-1, N;$$

$$\widehat{N-1}|+2|\widehat{N-2}, N; \widehat{N-2}, N|)|\hat{N}; \widehat{N-2}|+4(|\widehat{N-2},$$

$$N; \widehat{N-1}|+|\widehat{N-1}; \widehat{N-2}, N|)|\widehat{N-1}, N+1; \widehat{N-2}|+$$

$$|\hat{N}; \widehat{N-3}, N-1|=0 \tag{5.3.35}$$

利用(5.3.34a)消去(5.3.36)中含$\widehat{N-4}$的项并除以 2,得

$$|\widehat{N-1}; \widehat{N-1}|(4|\widehat{N-2}, N, N+1; \widehat{N-2}|+2|\hat{N}; \widehat{N-3},$$

$$N|-2|\widehat{N-1}, N+2; \widehat{N-2}|+2|\widehat{N-1}, N+1; \widehat{N-3},$$

$$N-1|)+(4|\widehat{N-2}, N+1; \widehat{N-1}|-2|\widehat{N-3}, N-1,$$

$$N; \widehat{N-1}|+2|\widehat{N-1}; \widehat{N-3}, N-1, N|+2|\widehat{N-2},$$

$$N; \widehat{N-2}, N)|\hat{N}; \widehat{N-2}|-2(|\widehat{N-2}, N; \widehat{N-1}|+|\widehat{N-1};$$

$$\widehat{N-2}, N|)(|\widehat{N-1}, N+1; \widehat{N-2}|+|\hat{N}; \widehat{N-3}, N-1|)$$

利用行列式性质 5 上式可划为

$$|\widehat{N-1}; \widehat{N-1}|(2|\widehat{N-1}, N+1; \widehat{N-3}, N-1|-2|\widehat{N-1},$$

$$N+2; \widehat{N-2}|)+|\hat{N}; \widehat{N-2}|(2|\widehat{N-2}, N; \widehat{N-2}, N|-2|$$

$$\widehat{N-3}, N-1, N; \widehat{N-1}|)-2|\widehat{N-1}, N+1; \widehat{N-2}||\widehat{N-1};$$

$$\widehat{N-2}, N|+2|\widehat{N-2}, N; \widehat{N-1}||\widehat{N-1}, N+1; \widehat{N-2}|$$

$$\tag{5.3.36}$$

利用(5.3.34b,c)消去(5.3.38)中含$\widehat{N-3}$的项,由行列式性质 5 可得
上式为零,即双 Wronskian 形式的 f 和 g 满足双线性方程(5.3.5a).

满足条件(5.3.28)的 ϕ 和 ψ 为

$$\phi_j = (t+k_j)^{N-1}\mathrm{e}^{-\rho_j}\,,\psi_j=(t+k_j)^{N-1}\mathrm{e}^{\rho_j} \qquad (5.3.37\text{a})$$

$$\phi_{N+j} = -\ \psi_j^*\,,\psi_{N+j}=\phi_j^*\,,\ (j=1,\ 2,\ \cdots,\ N) \qquad (5.3.37\text{b})$$

其中

$$\rho_j = \frac{\mathrm{i}x}{2(t+k_j)} + \rho_j^{(0)} \qquad (5.3.37\text{c})$$

k_j 和 $\rho_j^{(0)}$ 是复常数.

则当 $N=1$ 时，

$$f = \begin{vmatrix} \phi & \psi \\ -\psi^* & \phi^* \end{vmatrix},\ g=\begin{vmatrix} \psi & \partial_x\psi \\ \phi^* & \partial_x\phi^* \end{vmatrix} \qquad (5.3.38\text{a})$$

其中

$$\phi = \mathrm{e}^{-\rho},\ \psi=\mathrm{e}^{\rho},\ \rho=\frac{\mathrm{i}x}{2(t+k)}+\rho^{(0)}=K(t)x+\rho^{(0)}$$

$$(5.3.38\text{b})$$

$K(t)$ 是复函数，$\rho^{(0)}$ 是复常数. 则由（5.3.2）知一孤子解的模的平方为：

$$|q|^2 = (K(t)+K^*(t))^2\,\mathrm{sech}^2(\rho+\rho^*) \qquad (5.3.38\text{c})$$

§5.3.3　推广的双 Wronskian 解

定理 5.3　非等谱非线性 Schrödinger 方程的双线性方程（5.3.5）具有 $2N\times2N$ 阶推广的双 Wronskians 解

$$f = |\ \widehat{N-1};\ \widehat{N-1}\ |,\ g=2\,|\ \hat{N};\ \widehat{N-2}\ | \qquad (5.3.39)$$

$$f = f^*,\ g^*=2\,|\ \widehat{N-2};\ \hat{N}\ | \qquad (5.3.40)$$

其中 ϕ 和 ψ 满足

$$\phi_{j,\,x} = \sum_{s=1}^{j} \alpha_{js}\,\phi_s\,, \quad \psi_{j,\,x} = -\sum_{s=1}^{j} \alpha_{js}\,\psi_s\,, \quad (1 \leqslant j \leqslant 2N)$$

$$(5.3.41a)$$

$$\phi_{j,\,t} = 2i[x\phi_{j,\,xx} - (N-1)\,\phi_{j,\,x}]\,, \quad \psi_{j,\,t}$$

$$= -2i\,[x\psi_{j,\,xx} - (N-1)\,\psi_{j,\,x}]\,, (1 \leqslant j \leqslant 2N)$$

$$(5.3.41b)$$

α_{js} 是与 x 无关的函数.

证明　注意到所谓推广的双 Wronskian 解和普通的双 Wronskian 解是后者条件(5.3.28a)是前者的特殊情形(5.3.41a) [49,52]

$$\phi_{j,\,x} = -\alpha_{jj}\,\phi_j\,, \quad \psi_{j,\,x} = \alpha_{jj}\,\psi_j\,, \quad (1 \leqslant j \leqslant 2N)$$

$$(5.3.41a')$$

在定理 5.1 里的证明中,行列式的一些关系式是验证解的基础. 这里虽然用(5.3.41a)代替了(5.3.28a),利用行列式的性质 1 和 2 仍然有

$$\sum_{j=1}^{2N} \alpha_{jj} \mid \widehat{N-1};\ \widehat{N-1} \mid = \mid \widehat{N-2},\ N;\ \widehat{N-1} \mid - \mid \widehat{N-1};\ \widehat{N-2},\ N \mid$$

$$(5.3.42)$$

又由(5.3.41a)可得

$$\phi_{j,\,xx} = \alpha_{jj}^2\,\phi_j + \sum_{s=1}^{j-1}\beta_{js}\,\phi_s\,, \quad \psi_{j,\,xx} = \alpha_{jj}^2\,\psi_j + \sum_{s=1}^{j-1}\beta_{js}\,\psi_s\,, \quad (1 \leqslant j \leqslant 2N)$$

和

$$\sum_{j=1}^{2N} \alpha_{jj}^2 \mid \widehat{N-1};\ \widehat{N-1} \mid = -\mid \widehat{N-3},\ N-1,\ N;\ \widehat{N-1} \mid + \mid \widehat{N-2},$$

$$N+1;\ \widehat{N-1} \mid - \mid \widehat{N-1};\ \widehat{N-3},\ N-1,\ N \mid + \mid \widehat{N-1};$$

$$\widehat{N-2},\ N+1 \mid$$

$$(5.3.43)$$

利用

$$f\Big[\sum_{j=1}^{2N}\alpha_{jj}\Big(\sum_{j=1}^{2N}\alpha_{jj}\,f\Big)\Big]=\Big(\sum_{j=1}^{2N}\alpha_{jj}\,f\Big)^2 \tag{5.3.44}$$

可给出

$$f(\mid\widehat{N-3},\,N-1,\,N;\,\widehat{N-1}\mid+\mid\widehat{N-2},\,N+1;$$

$$\widehat{N-1}\mid-2\mid\widehat{N-2},\,N;\,\widehat{N-2},\,N\mid+\mid\widehat{N-1};$$

$$\widehat{N-3},\,N-1,\,N\mid+\mid\widehat{N-1};\,\widehat{N-2},\,N+1\mid)$$

$$=(\mid\widehat{N-2},\,N;\,\widehat{N-1}\mid-\mid\widehat{N-1};\,\widehat{N-2},\,N\mid)^2 \tag{5.3.45}$$

下面几个关系式是证明过程所需要的

$$f\Big(\sum_{j=1}^{2N}\alpha_{jj}^2\,g\Big)=g\Big(\sum_{j=1}^{2N}\alpha_{jj}^2\,f\Big) \tag{5.3.46}$$

$$f\Big(\sum_{j=1}^{2N}\alpha_{jj}\mid\widehat{N-2};\,\widehat{N-1},\,N+1\mid\Big)=\mid\widehat{N-2};\,\widehat{N-1},\,N+1\mid\Big(\sum_{j=1}^{2N}\alpha_{jj}\,f\Big)$$

$$\tag{5.3.47}$$

和

$$g\Big(\sum_{j=1}^{2N}\alpha_{jj}\mid\widehat{N-1};\,\widehat{N-2},\,N\mid\Big)=\mid\widehat{N-1};\,\widehat{N-2},\,N\mid\Big(\sum_{j=1}^{2N}\alpha_{jj}\,g\Big)$$

$$\tag{5.3.48}$$

利用以上关系式可类似于定理 5.1 给出证明.

为了得到更多的解,考虑下列函数

$$\widetilde{\phi}(k,\,t,\,x)=\mathrm{e}^{-\rho},\,\widetilde{\psi}(k,\,t,\,x)=\mathrm{e}^{\rho},\,\rho=\frac{\mathrm{i}x}{2(t+k)}+\rho^{(0)}$$

$$\tag{5.3.49}$$

k 和 $\rho^{(0)}$ 是复常数. 在 $\delta=0$ 作 Taylor 展开

$$\partial_x \widetilde{\phi}(k+\delta, t, x) = -\frac{i}{2(t+k+\delta)} \widetilde{\phi}(k+\delta, t, x)$$

$$(5.3.50)$$

比较 δ 的同次幂系数,有

$$Q_{s,x}(k) = \sum_{m=0}^{s} \alpha_{sm} Q_m(k), \quad s = 0, 1, \cdots \quad (5.3.51)$$

和

$$Q_{s,t}(k) = -2i[x Q_{s,xx}(k) - (N-1)Q_{s,x}(k)] \quad (5.3.52)$$

其中

$$\alpha_{sm} = (-1)^{s-m} \frac{s!}{m!} \frac{i}{2(t+k)^{s-m+1}} \quad (5.3.53)$$

$$Q_s(k) = Q_s(k, t, x) = \partial_k^s \widetilde{\rho}(k, t, x) \quad (5.3.54)$$

类似地,取

$$R_s(k) = R_s(k, t, x) = \partial_k^s e^{\rho} \quad (5.3.55)$$

可以得到

$$R_{s,x}(k) = \sum_{m=0}^{s} \alpha_{sm} R_s(k), \quad s = 0, 1, \cdots \quad (5.3.56)$$

和

$$R_{s,t}(k) = 2i[x R_{s,xx}(k) - (N-1)R_{s,x}(k)] \quad (5.3.57)$$

因此,如果取

$$\phi = (Q_0(k), Q_1(k), \cdots, Q_{N-1}(k); -R_0^*(k), -R_1^*(k), \cdots, -R_{N-1}^*(k))^T$$

$$(5.3.58a)$$

和

$$\psi = (R_0(k), R_1(k), \cdots, R_{N-1}(k); Q_0^*(k), Q_1^*(k), \cdots, Q_{N-1}^*(k))^{\mathrm{T}}$$

(5.3.58b)

更一般地,可以取

$$\phi = (Q_0(k_1), \cdots, Q_{h_1-1}(k_1); Q_0(k_2), \cdots, Q_{h_2-1}(k_2); \cdots;$$

$$Q_0(k_s), \cdots, Q_{h_s-1}(k_s); \phi_{s+1}, \cdots, \phi_{s+m}; -R_0^*(k_1), \cdots,$$

$$-R_{h_1-1}^*(k_1); -R_0^*(k_2), \cdots, -R_{h_2-1}^*(k_2); \cdots;$$

$$-R_0^*(k_s), \cdots, -R_{h_s-1}^*(k_s); -\psi_{s+1}^*, \cdots, -\psi_{s+m}^*)^{\mathrm{T}}$$

(5.3.59a)

和

$$\psi = (R_0(k_1), \cdots, R_{h_1-1}(k_1); R_0(k_2), \cdots, R_{h_2-1}(k_2); \cdots;$$

$$R_0(k_s), \cdots, R_{h_s-1}(k_s); \psi_{s+1}, \cdots, \psi_{s+m}; Q_0^*(k_1), \cdots,$$

$$Q_{h_1-1}^*(k_1); Q_0^*(k_2), \cdots, Q_{h_2-1}^*(k_2); \cdots;$$

$$Q_0^*(k_s), \cdots, Q_{h_s-1}^*(k_s); \phi_{s+1}^*, \cdots, \phi_{s+m}^*)^{\mathrm{T}}$$

(5.3.59b)

式中 $m + \sum_{j=1}^{s} h_j = N$, $h_j \geqslant 1$.

§5.3.4 解的性质

一、双线性解的性质

由(5.3.16)可以得到双线性形式的单孤子解的模平方为

$$|v|^2 = \left[\frac{\mathrm{Re}[k_1]}{[\mathrm{Im}[k_1]-t]^2 + \mathrm{Re}[k_1]^2}\right]^2 \mathrm{sech}^2\left[\frac{\mathrm{Re}[k_1]x}{[\mathrm{Im}[k_1]-t]^2 + \mathrm{Re}^2[k_1]} - \right.$$

$$\left. \ln 2\,\mathrm{Re}[k_1] + \mathrm{Re}[\xi_1^{(0)}]\right]$$

(5.3.60)

其振幅是 t 的四次有理函数,随时间变化;波峰的轨迹是抛物线:

$$x = -\left(\ln 2\,\mathrm{Re}[k_1] - \mathrm{Re}[\xi_1^{(0)}]\right)\frac{\left[\mathrm{Im}[k_1]-t\right]^2 + \mathrm{Re}^2[k_1]}{\mathrm{Re}[k_1]}$$

$$(5.3.61)$$

若 $\ln 2\,\mathrm{Re}[k_1] - \mathrm{Re}[\xi_1] \neq 0$，孤子沿抛物线传播，波峰在点 $(t,x) = (-\mathrm{Im}[k_1], -2\,\mathrm{Re}[k_1](\ln 2\,\mathrm{Re}[k_1]-\mathrm{Re}[\xi_1^{(0)}]))$，该点的振幅达到最大值 $\dfrac{1}{\mathrm{Re}^2[k_1(0)]}$.

波峰的速度（即孤子的速度）为

$$x'(t) = 2(\ln 2\,\mathrm{Re}[k_1] - \mathrm{Re}[\xi_1^{(0)}])\frac{\left[\mathrm{Im}[k_1]-t\right]}{\mathrm{Re}[k_1]} \quad (5.3.62)$$

当 $\mathrm{Re}[\xi_1^{(0)}] = \ln 2\,\mathrm{Re}[k_1]$，为稳态孤子；当 $\mathrm{Re}[\xi_1^{(0)}] \neq \ln 2\,\mathrm{Re}[k_1]$ 为运动孤子. 图 1(a). $k_1 = 3$，$\mathrm{Re}[\xi_1^{(0)}] = \ln 6$ 的稳态孤子，图 1(b). $k_1 = 3$，$\xi_1^{(0)} = -2$ 的运动孤子.

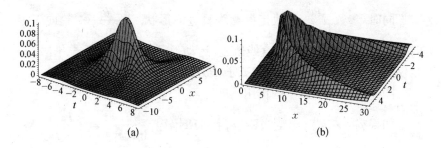

图 1 1-soliton 的形状和运动情况

(a) $k_1 = 3$，$\xi_1^{(0)} = \ln 6$ 的稳态孤子.

(b) $k_1 = 3$，$\xi_1^{(0)} = -2$ 的运动孤子.

二、Wronskian 解的性质

从 (5.3.38c)，我们可以给出一孤子解的另一种形式

$$|q|^2 = \left[\frac{\text{Im}[k_1]}{(\text{Re}[k_1]+t)^2 + \text{Im}^2[k_1]}\right]^2 \text{sech}^2\left[\frac{\text{Im}[k_1]x}{(\text{Re}[k_1]+t)^2 + \text{Im}^2[k_1]} + 2\rho_1^{(0)}\right]$$

(5. 3. 63)

振幅是 t 的四次有理函数,随时间变化,波峰的轨迹是抛物线:

$$x = -\frac{2\rho_1^{(0)}[(\text{Re}[k_1]+t)^2 + \text{Im}^2[k_1]]}{\text{Im}[k_1]}$$

(5. 3. 64)

如果 $\rho_1^{(0)} \neq 0$,孤子沿抛物线传播;波峰在点 $(t, x) = (-\text{Re}[k_1], -2\text{Im}[k_1]\rho_1^{(0)})$,该点的振幅达到最大值 $\frac{1}{\text{Im}^2[k_1]}$.

波峰的速度(即孤子的速度)为

$$x'(t) = -\frac{4\rho_1^{(0)}(\text{Re}[k_1]+t)}{\text{Im}[k_1]}$$

(5. 3. 65)

当 $\rho_1^{(0)} = 0$,得到稳态孤子;

同时,参数 $\frac{\rho_1^{(0)}}{\text{Im}[k_1]}$ 决定抛物线轨迹的形状. 当 $\frac{\rho_1^{(0)}}{\text{Im}[k_1]}$ 趋向于零时,抛物线轨迹渐渐变为直线轨迹,运动孤子变为稳态孤子. 图 2(a) 是仅由 k_1 的虚部决定的稳态孤子,(b) 是运动孤子.

图 3 给出了双孤子的作用.

如果 $\rho_1^{(0)} \cdot \rho_2^{(0)} \neq 0$,当 $\text{Re}[k_1] = \text{Re}[k_2]$,$\frac{\rho_1^{(0)}}{\text{Im}[k_1]} = \frac{\rho_2^{(0)}}{\text{Im}[k_2]}$ 是,给出有界双孤子. 特别地当 $\rho_1^{(0)} = \rho_2^{(0)} = 0$,我们给出稳态孤子. 在这种情况下,在轨迹的顶点 $x=0$ 处

$$|q|^2 = \frac{\{\text{Im}[k_1][(\text{Re}[k_2]+t)^2 + \text{Im}^2[k_2]] + \text{Im}[k_2][(\text{Re}[k_1]+t)^2 + \text{Im}^2[k_1]]\}^2}{[(\text{Re}[k_1]+t)^2 + \text{Im}^2[k_1]]^2[(\text{Re}[k_2]+t)^2 + \text{Im}^2[k_2]]^2}$$

(5. 3. 66)

即在稳态时,不存在周期相互作用;另外,由(5. 3. 66)可以得出:如果

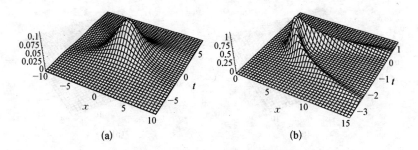

图 2   1-soliton 的形状和运动情况

(a) $k_1 = 3i$, $\rho_1^{(0)} = 0$ 的稳态孤子.

(b) $k_1 = 1 - i$, $\rho_1^{(0)} = 2$ 的运动孤子.

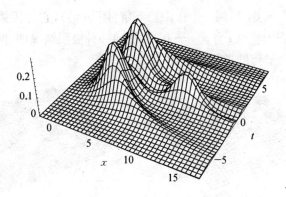

图 3 $k_1 = 2 - 2i$, $k_2 = -2 - 2i$, $\zeta_1^{(0)} = \zeta_2^{(0)} = 1$

双孤子的作用

$\text{Re}[k_1] = \text{Re}[k_2]$, $\text{Im}[k_1] = -\text{Im}[k_2]$, 给出零解. 图 4 给出了两个稳定的有界态. 图 4(b) 中 $|q|^2$ 在点 $(t, x) = (0, 0)$ 为零.

下面考察 (5.3.58a, b) 给出的推广的解

$$\phi = (Q_0(k_1), Q_1(k_1); -R_0^*(k_1), -R_1^*(k_1))^{\text{T}}$$
$$\psi = (R_0(k_1), R_1(k_1); Q_0^*(k_1), Q_1^*(k_1))^{\text{T}} \tag{5.3.67}$$

$Q_s(k_1)$ 和 $R_s(k_1)$ 由 (5.3.59) 给出.

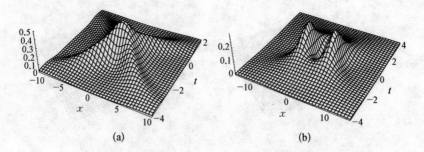

图 4 非等谱非线性 Schrödinger 方程的有界稳态

(a) $k_1 = 1 + 3i$, $k_2 = 1 - i$, $\rho_1^{(0)} = \rho_2^{(0)} = 0$.

(b) $k_1 = -1 - 2i$, $k_2 = 1 + 2i$, $\rho_1^{(0)} = \rho_2^{(0)} = 0$.

当 $\rho_1^{(0)} = 0$, 得到一个极限稳态解(图 5(a)), 它的传播行为和图 5(a) 的稳态解类似; 当 $\rho_1^{(0)} \neq 0$, 解描述一种极限情况即: 两个有界孤子沿着两个平行的抛物线运动(图 5(b)).

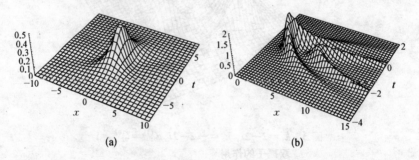

图 5 非等谱非线性 Schrödinger 方程推广的 Wronskian 解的形状与运动情况

(a) $k_1 = 3i$, $\rho_1^{(0)} = 0$ 稳态极限态.

(b) $k_1 = 1 - i$, $\rho_1^{(0)} = 2$ 运动极限态.

§5.4 KdV 系统的 τ 方程的解

考虑 KdV 系统的 τ 方程

$$u_t = 3t(u_{xxx} + 6uu_x) + xu_x + 2u \qquad (5.4.1)$$

引入变换

$$u = 2(\ln f)_{xx} \qquad (5.4.2)$$

代入方程(5.4.1)得双线性方程

$$(D_x D_t - 3tD_x^4 - xD_x^2)f \cdot f - 2ff_x \qquad (5.4.3)$$

将 f 摄动展开

$$f = 1 + \varepsilon f^{(1)} + \varepsilon^2 f^{(2)} + \varepsilon^3 f^{(3)} + \cdots \qquad (5.4.4)$$

代入(5.4.3),比较 ε 的同次幂系数,得

$$f_{xt}^{(1)} - 3tf_{xxxx}^{(1)} - xf_{xx}^{(1)} - f_x^{(1)} = 0 \qquad (5.4.5a)$$

$$f_{xt}^{(2)} - 3tf_{xxxx}^{(2)} - xf_{xx}^{(2)} - f_x^{(1)}$$

$$= -\frac{1}{2}(D_x D_t - 3tD_x^4 - xD_x^2)f^{(1)} \cdot f^{(1)} + f^{(1)} f_x^{(1)}$$

$$(5.4.5b)$$

$$f_{xt}^{(3)} - 3tf_{xxxx}^{(3)} - xf_{xx}^{(3)} - f_x^{(3)} = -(D_x D_t - 3tD_x^4 - xD_x^2)f^{(1)} \cdot f^{(2)} +$$

$$(f^{(1)} f^{(2)})_x \qquad (5.4.5c)$$

$$\cdots$$

设

$$f^{(1)} = e^{\xi_1}, \ \xi_1 = k_1(t)x + \omega_1(t) + \xi_1^{(0)} \qquad (5.4.6)$$

代入(5.4.5a),给出

$$k_1'(t) + xk_1(t)k_1'(t) + k_1(t)\omega_1'(t) - 3tk_1^4(t) - xk_1^2(t) - k_1(t) = 0$$

$$(5.4.7)$$

因此,有

$$k_1'(t) = k_1(t), \ \omega_1'(t) = 3tk_1^3(t) \qquad (5.4.8a,b)$$

解此方程得

$$k_1(t) = k_1 \mathrm{e}^t \,,\; \omega_1(t) = k_1^3 t \mathrm{e}^{3t} - \frac{1}{3} k_1^3 \mathrm{e}^{3t} \qquad (5.4.9\mathrm{a,b})$$

式中 k_1 是常数.

将 $f^{(1)}$ 代入 $(5.4.5\mathrm{b,c})$，可依次推出

$$f^{(j)} = 0 \,,\; j = 2,\, 3,\, \cdots \qquad (5.4.10)$$

令 $\varepsilon = 1$，得 $(5.4.1)$ 的单孤子解

$$u = \frac{k_1^2 \mathrm{e}^{2t}}{2} \,\mathrm{sech}^2 \frac{k_1 x \mathrm{e}^t + k_1^3 t \mathrm{e}^{3t} - \frac{1}{3} k_1^3 \mathrm{e}^{3t} + \xi_1^{(0)}}{2} \qquad (5.4.11)$$

若取

$$f^{(1)} = \mathrm{e}^{\xi_1} + \mathrm{e}^{\xi_2} \,,\; \xi_j = k_j(t)x + \omega_j(t) + \xi_j^{(0)} \qquad (5.4.12\mathrm{a})$$

$$k_j'(t) = k_j(t) \,,\; \omega_j(t)' = 3t k_j(t) \,,\; j = 1,\, 2 \qquad (5.4.12\mathrm{b})$$

代入 $(5.4.5\mathrm{b})$，给出

$$f^{(2)} = \mathrm{e}^{\xi_1 + \xi_2 + \theta_{12}} \,,\; \mathrm{e}^{\theta_{12}} = \frac{(k_1 - k_2)^2}{(k_1 + k_2)^2} \qquad (5.4.13)$$

将 $f^{(1)}$，$f^{(2)}$ 代入 $(5.4.5\mathrm{c})$ 可依次给出

$$f^{(j)} = 0 \,,\; j = 3,\, 4,\, \cdots \qquad (5.4.14)$$

因此 $(5.4.1)$ 的二孤子解为

$$u = 2\ln \left(1 + \mathrm{e}^{\xi_1} + \mathrm{e}^{\xi_2} + \mathrm{e}^{\xi_1 + \xi_2 + \theta_{12}}\right)_{xx} \qquad (5.4.15)$$

若取

$$f^{(1)} = \sum_{j=1}^{n} \mathrm{e}^{\xi_j} \,,\; \xi_j = k_j(t)x + \omega_j(t) + \xi_j^{(0)} \qquad (5.4.16)$$

可得 τ 方程的 N 孤子解

$$u = 2\Big[\ln\Big(\sum_{\mu=0,1} e^{\sum_{j=1}^{n}\mu_j\xi_j+\sum_{1\leqslant j<l}^{n}\mu_j\mu_l\theta_{jl}}\Big)\Big]_{xx} \qquad (5.4.17)$$

式中 $e^{\theta_{jl}} = \dfrac{(k_j-k_l)^2}{(k_j+k_l)^2}.$

§5.5 mKdV 系统的 τ 方程的解

考虑 mKdV 方程的 τ 方程

$$v_t = 3t(v_{xxx} + 6v^2 v_x) + (xv)_x \qquad (5.5.1)$$

引入变换

$$v = i\Big(\ln\frac{f^*}{f}\Big)_x \qquad (5.5.2)$$

代入方程(5.5.1)给出

$$f_t^* f - f^* f_t = 3t\{f_{xxx}^* f - f^* f_{xxx} + \frac{3}{f^* f}[f^* f_x(f^* f_{xx} - 2f_x^* f_x) -$$

$$ff_x^*(f_{xx}^* f - 2f_x^* f_x)]\} + x(f_x^* f - f^* f_x) \qquad (5.5.3a)$$

即

$$D_t f^* \cdot f = 3t\Big[D_x^3 f^* \cdot f + \frac{3}{f^* f}(D_x f^* \cdot f)(D_x^2 f^* \cdot f)\Big] + xD_x f^* \cdot f$$

$$(5.5.3b)$$

可给出(5.5.1)的双线性方程为

$$(D_t - 3tD_x^3 - xD_x)f^* \cdot f = 0 \qquad (5.5.4a)$$

$$D_x^2 f^* \cdot f = 0 \qquad (5.5.4b)$$

将 f 摄动展开

$$f = 1 + \varepsilon f^{(1)} + \varepsilon^2 f^{(2)} + \varepsilon^3 f^{(3)} + \cdots \qquad (5.5.5)$$

比较 ε 的同次幂系数给出

$$(f_1^{(1)*} - f^{(1)})_t - 3t(f^{(1)*} - f^{(1)})_{xxx} - x(f^{(1)*} - f^{(1)})_x = 0 \tag{5.5.6a}$$

$$(f^{(2)*} - f^{(2)})_t - 3t(f^{(2)*} - f^{(2)})_{xxx} - x(f^{(2)*} - f^{(2)})_x$$
$$= -(D_t - 3tD_x^3 - xD_x)f^{(1)*} \cdot f^{(1)} \tag{5.5.6b}$$

$$(f^{(3)*} - f^{(3)})_t - 3t(f^{(3)*} - f^{(3)})_{xxx} - x(f^{(3)*} - f^{(3)})_x$$
$$= (D_t - 3tD_x^3 - xD_x)(f^{(2)*} \cdot f^{(1)} + f^{(1)*} \cdot f^{(2)}) \tag{5.5.6c}$$

$$\cdots$$

$$(f^{(1)*} + f^{(1)})_{xx} = 0 \tag{5.5.7a}$$

$$(f^{(2)*} + f^{(2)})_{xx} = -D_x^2 f^{(1)*} \cdot f^{(1)} \tag{5.5.7b}$$

$$(f^{(3)*} + f^{(3)})_{xx} = -D_x^2(f^{(2)*} \cdot f^{(1)} + f^{(1)*} \cdot f^{(2)}) \tag{5.5.7c}$$

$$\cdots$$

若取

$$f^{(1)} = \sum_{j=1}^{n} e^{\xi_j + \frac{1}{2}i\pi}, \quad \xi_j = k_j(t)x + \omega_j(t) + \xi_j^{(0)} \tag{5.5.8}$$

则,有

$$f = \sum_{\mu=0,1} e^{\sum_{j=1}^{N} \mu_j(\xi_j + \frac{\pi}{2}i) + \sum_{1 \leqslant j < l}^{N} \mu_j \mu_l A_{jl}} \tag{5.5.9}$$

式中

$$k_j(t) = k_j e^t, \quad \omega_j(t) = k_j^3 t e^{3t} - \frac{1}{3}k_j^3 e^{3t}, \quad k_j \text{ 是常数}$$

$$e^{A_{jl}} = \left[\frac{k_l - k_j}{k_l + k_j}\right]^2$$

对 $\mu = 0, 1$ 求表示 $\mu_j (j = 1, 2, \cdots)$，取 0 或 1 的所有可能之和.

因此 τ 方程(5.5.1)的 n 孤子解为

$$v = \mathrm{i} \left[\ln \frac{\sum\limits_{\mu = 0,1} \mathrm{e}^{\sum\limits_{j=1}^{N} \mu_j (\xi_j - \frac{\pi}{2}\mathrm{i}) + \sum\limits_{1 \leqslant j < l}^{N} \mu_j \mu_l A_{jl}}}{\sum\limits_{\mu = 0,1} \mathrm{e}^{\sum\limits_{j=1}^{N} \mu_j (\xi_j + \frac{\pi}{2}\mathrm{i}) + \sum\limits_{1 \leqslant j < l}^{N} \mu_j \mu_l A_{jl}}} \right]_x \tag{5.5.10}$$

§5.6 非线性 Schrödinger 系统的 τ 方程的解

考虑非线性 Schrödinger 方程的 τ 方程

$$v_t = -2\mathrm{i}t(v_{xx} + 2v \mid v \mid^2) + (xv)_x \tag{5.6.1}$$

引入变换

$$v = \frac{g}{f}, \quad f \text{ 是实函数} \tag{5.6.2}$$

则有

$$D_t g \cdot f = -2\mathrm{i}t(D_x^2 g \cdot f - \frac{g}{f}(D_x^2 f \cdot f - 2gg^*)) + fg + xD_x g \cdot f = 0 \tag{5.6.3}$$

可得双线性方程

$$(\mathrm{i}D_t - \mathrm{i}xD_x - 2tD_x^2)g \cdot f - \mathrm{i}fg = 0 \tag{5.6.4a}$$

$$D_x^2 f \cdot f = 2gg^* \tag{5.6.4b}$$

将 f, g 作摄动展开

$$f = 1 + \varepsilon^2 f^{(2)} + \varepsilon^4 f^{(4)} + \varepsilon^6 f^{(6)} + \cdots \tag{5.6.5a}$$

$$g = \varepsilon g^{(1)} + \varepsilon^3 g^{(3)} + \varepsilon^5 g^{(5)} + \cdots \tag{5.6.5b}$$

比较 ε 的同次幂系数给出

$$\mathrm{i}g_t^{(1)} - \mathrm{i}xg_x^{(1)} - 2tg_{xx}^{(1)} - \mathrm{i}g^{(1)} = 0 \qquad (5.6.6a)$$

$$\mathrm{i}g_t^{(3)} - \mathrm{i}xg_x^{(3)} - 2tg_{xx}^{(3)} - \mathrm{i}g^{(3)}$$

$$= -(\mathrm{i}D_t - \mathrm{i}xD_x - 2tD_x^2)g^{(1)} \cdot f^{(2)} + \mathrm{i}g^{(1)}f^{(2)} \qquad (5.6.6b)$$

$$\mathrm{i}g_t^{(5)} - \mathrm{i}xg_x^{(5)} - 2tg_{xx}^{(5)} - \mathrm{i}g^{(5)} = -(\mathrm{i}D_t - \mathrm{i}xD_x - 2tD_x^2)(g^{(3)} \cdot f^{(2)} +$$

$$g^{(1)}f^{(4)}) + \mathrm{i}g^{(3)}f^{(2)} + \mathrm{i}g^{(1)}f^{(4)}$$

$$(5.6.6c)$$

$$\cdots$$

$$2f_{xx}^{(2)} = 2g^{(1)}g^{(1)*} \qquad (5.6.7a)$$

$$2f_{xx}^{(4)} = -D_x^2 f^{(2)} \cdot f^{(2)} + 2g^{(1)}g^{(3)*} + 2g^{(1)*}g^{(3)}$$

$$(5.6.7b)$$

$$2f_{xx}^{(6)} = -2D_x^2 f^{(2)} \cdot f^{(4)} + 2g^{(1)}g^{(5)*} + 2g^{(3)}g^{(3)*} + 2g^{(5)}g^{(1)*}$$

$$(5.6.7c)$$

由$(5.6.6a)$，可取

$$g^{(1)} = \mathrm{e}^{\xi_1}, \ \xi_1 = k_1(t)x + \omega_1(t) + \xi_1^{(0)}, \ \xi_1^{(0)} \text{是常数}$$

$$k_1(t) = k_1 \mathrm{e}^t, \ \omega_1(t) = t - \mathrm{i}tk_1^2 \mathrm{e}^{2t} + \frac{1}{2}\mathrm{i}k_1^2 \mathrm{e}^{2t}, \ k_1 \text{是常数}$$

$$(5.6.8)$$

代入$(5.6.7a)$得

$$f^{(2)} = \frac{1}{(k_1 + k_1^*)^2 \mathrm{e}^{2t}} \mathrm{e}^{\xi_1 + \xi_1^*} \qquad (5.6.9)$$

将$g^{(1)}$，$f^{(2)}$代入$(5.6.6b)$，$(5.6.7b)$，可得$g^{(3)} = f^{(4)} = 0$，取$\varepsilon = 1$，
有

$$f = 1 + \frac{1}{(k_1 + k_1^*)^2 e^{2t}} e^{\xi_1 + \xi_1^*} , \quad g = e^{\xi_1} \qquad (5.6.10)$$

则，非线性 Schrödinger 的 τ 方程(5.6.1)的单孤子解为

$$v = \frac{e^{\xi_1}}{1 + \dfrac{1}{(k_1 + k_1^*)^2 e^{2t}} e^{\xi_1 + \xi_1^*}}$$

$$= \frac{e^{k_1 x e^t + t - itk_1^2 e^{2t} + \frac{1}{2} ik_1^2 e^{2t} + \xi_1^{(0)}}}{1 + \dfrac{1}{(k_1 + k_1^*)^2} e^{x(k_1 + k_1^*) e^t - i(k_1^2 - k_1^{*2})te^{2t} + \frac{1}{2} i(k_1^2 - k_1^{*2})e^{2t} + \xi_1^{(0)} + \xi_1^{*(0)}}}$$

$$\qquad (5.6.11)$$

一般地，如果取

$$g^{(1)} = \sum_{j=1}^n e^{\xi_j} , \quad \xi_j = k_j(t)x + \omega_j(t) + \xi_j^{(0)} , \quad \xi_j^{(0)} \text{ 是常数}$$

$$k_j(t) = k_j e^t , \quad \omega_j(t) = t - itk_j^2 e^{2t} + \frac{1}{2} ik_j^2 e^{2t} , \quad k_j \text{ 是常数}$$

$$\qquad (5.6.12)$$

可得

$$f = \sum_{\mu=0,1} A_1(\mu) e^{\sum_{j=1}^{2n} \mu_j \xi_j + \sum_{1 \le j < l}^{2n} \mu_j \mu_l \theta_{jl}} \qquad (5.6.13a)$$

$$g = \sum_{\mu=0,1} A_2(\mu) e^{\sum_{j=1}^{2n} \mu_j \xi_j + \sum_{1 \le j < l}^{2n} \mu_j \mu_l \theta_{jl}} \qquad (5.6.13b)$$

其中

$$\xi_{n+j} = \xi_j^* , \quad (j = 1, 2, \cdots) \qquad (5.6.14a)$$

$$e^{\theta_{j(n+l)}} = \frac{1}{(k_j(t) + k_l^*(t))^2} , \quad (j, l = 1, 2, \cdots, n) \qquad (5.6.14b)$$

135 ◄

$$e^{\theta_{jl}} = (k_j(t) - k_l(t))^2, \ (j < l = 2, 3, \cdots, n)$$

(5. 6. 14c)

$$e^{\theta_{(n+j)(n+l)}} = (k_j^*(t) - k_l^*(t))^2$$ (5. 6. 14d)

而 $A_1(\mu)$ 与 $A_2(\mu)$ 表示当 $\mu_j(j = 1, 2, \cdots, n)$ 取所有可能的 0 或 1 时,还需要分别满足条件

$$\sum_{j=1}^{n} \mu_j = \sum_{j=1}^{n} \mu_{n+j}, \ \sum_{j=1}^{n} \mu_j = \sum_{j=1}^{n} \mu_{n+j} + 1$$ (5. 6. 15)

则,非线性 Schrödinger 的 τ 方程(5.6.1)的 N 孤子解

$$v = \frac{\sum\limits_{\mu=0,1} A_2(\mu) e^{\sum\limits_{j=1}^{2n} \mu_j \xi_j + \sum\limits_{1 \leqslant j < l}^{2n} \mu_j \mu_l \theta_{jl}}}{\sum\limits_{\mu=0,1} A_1(\mu) e^{\sum\limits_{j=1}^{2n} \mu_j \xi_j + \sum\limits_{1 \leqslant j < l}^{2n} \mu_j \mu_l \theta_{jl}}}$$ (5. 6. 16)

§5.7 sine-Gordon 系统的 τ 方程的解

本节考虑 sine-Gordon 方程的 τ 方程的解

$$u_{xt} = t\sin u + (\cos u \partial^{-1} \cos u \ \partial^{-1} + \sin u \partial^{-1} \sin u \partial^{-1})(xu_x)_x$$

(5. 7. 1)

引入变换

$$u = 2\mathrm{i}\ln \frac{f^*}{f}$$ (5. 7. 2)

代入(5.7.1),有

$$\mathrm{i}\frac{1}{f^{*2}}D_x D_t f^* \cdot f^* - \mathrm{i}\frac{1}{f^2}D_x D_t f \cdot f = \frac{1}{2\mathrm{i}}t\left(\frac{f^2-f^{*2}}{f^{*2}} - \frac{f^{*2}-f^2}{f^2}\right) +$$

$$\frac{x}{2\mathrm{i}}\left(\frac{f^2}{f^{*2}} - \frac{f^{*2}}{f^2}\right) + \frac{1}{2\mathrm{i}}\frac{f^2}{f^{*2}}\partial^{-1}\frac{f^{*2}-f^2}{f^2} - \frac{1}{2\mathrm{i}}\frac{f^{*2}}{f^2}\partial^{-1}\frac{f^2-f^{*2}}{f^{*2}}$$

(5. 7. 3)

做辅助函数 $g(t, x)$，使得

$$\frac{f^2 - f^{*2}}{f^{*2}} = \left(\frac{g}{f^*}\right)_x \qquad (5.7.4)$$

则，得双线性方程

$$D_x D_t f \cdot f - \frac{1}{2}t(f^2 - f^{*2}) - \frac{x}{2}(f^2 - f^{*2}) + \frac{1}{2}f^* g = 0 \qquad (5.7.5a)$$

$$D_x g \cdot f^* + f^{*2} - f^2 = 0 \qquad (5.7.5b)$$

将 f 和 g 做摄动展开

$$f = 1 + f^{(1)}\varepsilon + f^{(2)}\varepsilon^2 + f^{(3)}\varepsilon^3 + \cdots \qquad (5.7.6a)$$

$$g = g^{(1)}\varepsilon + g^{(2)}\varepsilon^2 + g^{(3)}\varepsilon^3 + \cdots \qquad (5.7.6b)$$

代入(5.7.5)，比较 ε 的同次幂系数给出

$$2f_{xt}^{(1)} - t(f^{(1)} - f^{*(1)}) - x(f^{(1)} - f^{*(1)}) + \frac{1}{2}g^{(1)} = 0 \qquad (5.7.7a)$$

$$2f_{xt}^{(2)} - t(f^{(2)} - f^{*(2)}) - x(f^{(2)} - f^{*(2)}) + \frac{1}{2}g^{(2)}$$

$$= \frac{t}{2}(f^{(1)}f^{(1)} - f^{*(1)}f^{*(1)}) + \frac{x}{2}(f^{(1)}f^{(1)} - f^{*(1)}f^{*(1)}) -$$

$$\frac{1}{2}f^{*(1)}g^{(1)} - D_x D_t f^{(1)} \cdot f^{(1)} \qquad (5.7.7b)$$

$$\cdots$$

$$g_x^{(1)} + 2(f^{*(1)} - f^{(1)}) = 0 \qquad (5.7.8a)$$

$$g_x^{(2)} - 2f^{(2)} + 2f^{*(2)} = -D_x g^{(1)} \cdot f^{(1)*} + f^{(1)}f^{(1)} - f^{*(1)}f^{*(1)} \qquad (5.7.8b)$$

···

令

$$f^{(1)} = e^{k_1(t)x + \omega_1(t) + \xi_1^{(0)} + \frac{i\pi}{2}} = e^{\xi_1 + \frac{i\pi}{2}} \tag{5.7.9}$$

由(5.7.8a)得

$$g^{(1)} = \frac{4}{k_1(t)} e^{\xi_1 + \frac{i\pi}{2}} \tag{5.7.10}$$

将(5.7.9)、(5.7.10)代入(5.7.7a),给出

$$k_1' + xk_1k_1' + k_1\omega_1' - t - x + \frac{1}{k_1} = 0 \tag{5.7.11}$$

因此有

$$k_1' = \frac{1}{k_1}, \quad \omega_1' = \frac{t}{k_1} - 2\frac{1}{k_1^2} \tag{5.7.12a,b}$$

解此方程,得

$$k_1(t) = \sqrt{2t + c_1}, \quad \omega_1(t) = t\sqrt{2t + c_1} - \frac{1}{3}(2t + c_1)^{\frac{3}{2}} -$$

$$\ln(2t + c_1), \quad c_1 \text{ 是常数} \tag{5.7.13a, b}$$

将 $f^{(1)}$, $g^{(1)}$ 代入(5.7.7b),(5.7.8b)可依次推出

$$f^{(j)} = 0, \quad g^{(j)} = 0, \quad j = 2, 3, \cdots \tag{5.7.14}$$

f, g 的摄动展开可截断为有限项.

令 $\varepsilon = 1$, 得(5.7.1)的单孤子解为

$$v = 2i \ln \frac{1 + e^{x\sqrt{2t+c_1} + t\sqrt{2t+c_1} - \frac{1}{3}(2t+c_1)^{\frac{3}{2}} - \ln(2t+c_1) - \frac{1}{2}i\pi}}{1 + e^{x\sqrt{2t+c_1} + t\sqrt{2t+c_1} - \frac{1}{3}(2t+c_1)^{\frac{3}{2}} - \ln(2t+c_1) + \frac{1}{2}i\pi}} \tag{5.7.15}$$

若取

$$f^{(1)} = \sum_{j=1}^{N} e^{\xi_j + \frac{\pi}{2} i}, \ \xi_j = k_j(t)x + \omega_j(t) + \xi_j^{(0)}, \ \xi_j^{(0)} \text{是实常数}$$

$$(5.7.16)$$

$$k_j(t) = \sqrt{c_j + 2t}, \ c_j^{(0)} \text{是实常数} \qquad (5.7.17a)$$

$$\omega_j(t) = t\sqrt{2t + c_j} - \frac{1}{3}(2t + c_j)^{\frac{3}{2}} - \ln(c_j + 2t)$$

$$(5.7.17b)$$

代入(5.7.7)、(5.7.8)可以得到

$$f = \sum_{\varepsilon = 0,1} \exp\left[\sum_{j=1}^{N} \varepsilon_j(\xi_j + \frac{\pi}{2}i) + \sum_{1 \leqslant j < l}^{N} \varepsilon_j \varepsilon_l A_{jl} \right] \quad (5.7.18)$$

式中 $e^{A_{jl}} = \left[\dfrac{k_l(t) - k_j(t)}{k_l(t) + k_j(t)} \right]^2$.

则，τ 方程(5.7.1)的 N 孤子解为

$$v = 2i\ln \frac{\sum\limits_{\varepsilon=0,1} \exp\left[\sum\limits_{j=1}^{N} \varepsilon_j(\xi_j - \frac{\pi}{2}i) + \sum\limits_{1 \leqslant j < l}^{N} \varepsilon_j \varepsilon_l A_{jl} \right]}{\sum\limits_{\varepsilon=0,1} \exp\left[\sum\limits_{j=1}^{N} \varepsilon_j(\xi_j + \frac{\pi}{2}i) + \sum\limits_{1 \leqslant j < l}^{N} \varepsilon_j \varepsilon_l A_{jl} \right]} (5.7.19)$$

第六章 一阶非等谱方程族与等谱方程族之间规范变换

§6.1 规范变换简介

设给定两个线性问题

$$\phi_x = M(u, \eta)\phi, \ \phi_t = N(u, \eta)\phi \qquad (6.1.1a, \ b)$$

$$\psi_x = M'(v, \eta)\psi, \ \psi_t = N'(v, \eta)\psi \qquad (6.1.2a, \ b)$$

其中 $\phi = (\phi_1, \phi_2, \cdots, \phi_n)^T$ 与 $\psi = (\psi_1, \psi_2, \cdots, \psi_n)^T$ 是 n 维本征向量函数，$M(u, \eta)$ 与 $N(u, \eta)$ 是依赖于向量位势 $u = (u_1, u_2, \cdots, u_n)$ 与谱参数 η 的 n 阶矩阵，而 $M'(v, \eta)$ 与 $N'(v, \eta)$ 是依赖于向量位势 $v = (v_1, v_2, \cdots, v_n)$ 与同一谱参数 η 的 n 阶矩阵. 如果存在 n 阶可逆矩阵 $T(u, v, \eta)$，使得本征向量函数的变换

$$\psi = T(u, v, \eta)\phi \qquad (6.1.3)$$

将线性问题(6.1.2)化为线性问题(6.1.1)，则称此变换为规范变换，且线性问题(6.1.1)与(6.1.2)称为在规范变换意义下等价.

将变换(6.1.3)代入 (6.1.2)容易得

$$T_x = M'T - TM \qquad (6.1.4a)$$

$$T_t = N'T - TN \qquad (6.1.4b)$$

这是确定变换矩阵 T 的一对方程. 如果线性问题 (6.1.1)与(6.1.2)各自相容,且为规范等价,则(6.1.4)是相容的. 事实上,因为

$$T_{xt} = (M'_t + M'N')T - M'TN - N'TM - T(M_t - NM)$$

$$\tag{6.1.5a}$$

$$T_{tx} = (N'_x + N'M')T - N'TM - M'TN - T(N_x - MN)$$

$$\tag{6.1.5b}$$

欲使 $T_{xt} = T_{tx}$，则必有

$$(M'_t - N'_x + [M', N'])T = T(M_t - N_x + [M, N])$$

$$\tag{6.1.6}$$

但 n 阶矩阵 M'、N' 与 M、N 分别满足零曲率方程，可见 (6.1.6)是一恒等式. 这就证明了所要的结论.

§6.2 一阶非等谱方程族与等谱方程族

为了叙述方便，等谱和非等谱方程族的谱问题和位势函数用不同符号表示.

以 AKNS 系统为例，考虑等谱问题

$$\phi_x = \begin{pmatrix} -\eta & q \\ r & \eta \end{pmatrix}\phi = M\phi \tag{6.2.1a}$$

$$\phi_t = \begin{pmatrix} A & B \\ C & -A \end{pmatrix}\phi = N\phi, \ \phi = \begin{bmatrix} \phi_1 \\ \phi_2 \end{bmatrix} \tag{6.2.1b}$$

和非等谱问题

$$\psi_x = \begin{pmatrix} -k & u \\ v & k \end{pmatrix}\psi = M'\psi \tag{6.2.2a}$$

$$\psi_t = \begin{pmatrix} A' & B' \\ C' & -A' \end{pmatrix}\psi = N'\psi, \ \psi = \begin{bmatrix} \psi_1 \\ \psi_2 \end{bmatrix} \tag{6.2.2b}$$

其中 $q = q(t, x)$，$r = r(t, x)$，$u = u(t, x)$，$v = v(t, x)$ 是光滑的

位势, k, η 是谱参数, 而 A, B, C 是变量 t, x, 位势 q, r 和谱参数 η 的待定函数; 而 A', B', C' 是变量 t, x, 位势 u, v 和谱参数 k 的待定函数. 谱参数满足

$$\eta_t = 0, \quad k_t = -\alpha k + \beta \qquad (6.2.3)$$

对 $k_t = \alpha k + \beta$, 关于 t 从 0 到 t 积分, 取 $k(0) = \eta$, 得

$$k(t) = \eta\, \mathrm{e}^{\alpha t} + \frac{\beta}{\alpha}(\mathrm{e}^{\alpha t} - 1) = \eta\, \mathrm{e}^{\alpha t} + \frac{b(t)}{2} \qquad (6.2.4)$$

代入非等谱问题 $(6.2.2)$, 化为

$$\psi_x = \begin{bmatrix} -\eta\, \mathrm{e}^{\alpha t} - \dfrac{\beta}{\alpha}(\mathrm{e}^{\alpha t} - 1) & u \\[2mm] v & \eta\, \mathrm{e}^{\alpha t} + \dfrac{\beta}{\alpha}(\mathrm{e}^{\alpha t} - 1) \end{bmatrix} \psi = M'' \psi$$

$$(6.2.5\mathrm{a})$$

$$\psi_t = \begin{pmatrix} A'' & B'' \\ C'' & -A'' \end{pmatrix} \psi = N'' \psi, \quad \psi = \begin{pmatrix} \psi_1 \\ \psi_2 \end{pmatrix} \qquad (6.2.5\mathrm{b})$$

如果给定边值条件

$$N\,\big|_{(q,\,r)=(0,\,0)} = \begin{bmatrix} -\dfrac{1}{2}(2\eta)^n & 0 \\[2mm] 0 & \dfrac{1}{2}(2\eta)^n \end{bmatrix} \qquad (6.2.6\mathrm{a})$$

和

$$N''\,\big|_{(u,\,v)=(0,\,0)} = \begin{bmatrix} -\alpha\eta\, x\, \mathrm{e}^{\alpha t} - \beta x\, \mathrm{e}^{\alpha t} - \dfrac{1}{2}(2\eta)^n & 0 \\[2mm] 0 & \alpha\eta\, x\, \mathrm{e}^{\alpha t} + \beta x\, \mathrm{e}^{\alpha t} + \dfrac{1}{2}(2\eta)^n \end{bmatrix}$$

$$(6.2.6\mathrm{b})$$

由(6.2.1a，b)和(6.2.5a，b)所满足的零曲率方程及系数矩阵 N 及 N'' 所满足边值条件可给出 AKNS 等谱方程族

$$\begin{pmatrix} q \\ r \end{pmatrix}_t = L^n \begin{pmatrix} -q \\ r \end{pmatrix} \tag{6.2.7}$$

和一阶 AKNS 非等谱方程族

$$\begin{pmatrix} u \\ v \end{pmatrix}_t = e^{-nat}(L'-b(t))^n \begin{pmatrix} -u \\ v \end{pmatrix} + \alpha \begin{pmatrix} xu \\ xv \end{pmatrix}_x + 2\beta \begin{pmatrix} -xu \\ xv \end{pmatrix} \tag{6.2.8}$$

式中 $L = \sigma\partial + 2\begin{pmatrix} q \\ -r \end{pmatrix}\partial^{-1}(r, q)$，$L' = \sigma\partial + 2\begin{pmatrix} u \\ -v \end{pmatrix}\partial^{-1}(v, u)$ $\sigma = \begin{pmatrix} -1 & 0 \\ 0 & 1 \end{pmatrix}$，$b(t) = 2\dfrac{\beta}{\alpha}(e^{at}-1)$.

§6.3 一阶非等谱方程族和等谱方程族之间的关系

一般地，在谱参数关于时间 t 的导数是谱参数的一次线性函数时，非等谱问题和等谱问题之间存在规范变换，所联系的一阶非等谱方程族和等谱方程族之间存在转换算子.

对谱问题(6.2.5)，首先做关于 x 的自变量变换

$$x = e^{-at}y \tag{6.3.1}$$

则谱问题(6.2.5)化为

$$\psi_y = (-\eta U + V)\psi \tag{6.3.2a}$$

$$U = \begin{pmatrix} 1 & 0 \\ 0 & -1 \end{pmatrix}, \quad V = \begin{pmatrix} -\dfrac{\beta}{\alpha}(1-e^{-at}) & ue^{-at} \\ ve^{-at} & \dfrac{\beta}{\alpha}(1-e^{-at}) \end{pmatrix}$$

$$\psi_t = \begin{bmatrix} \alpha\eta y + \beta y(1-\mathrm{e}^{-at}) + A'' & -\alpha y\mathrm{e}^{-at}u + B'' \\ -\alpha y\mathrm{e}^{-at}v + C'' & -\alpha\eta y - \beta y(1-\mathrm{e}^{-at}) - A'' \end{bmatrix}\psi$$

(6.3.2b)

为叙述的方便,重写谱问题(6.2.1)

$$\phi_y = (\eta\sigma + P)\phi, \quad P = \begin{pmatrix} 0 & q \\ r & 0 \end{pmatrix}$$

(6.3.3)

相应的等谱族(6.2)中的递推算子 $L = \sigma\,\partial_y + 2\begin{pmatrix} q \\ -r \end{pmatrix}\partial^{-1}(r, q)$.

寻找行列式值为 1 的不含 η 的本征函数变换 T,使得本征函数变换

$$\psi = T\phi$$

(6.3.4)

将(6.3.2)化为(6.3.3).

设变换矩阵 T 的为

$$T = \begin{bmatrix} T_1 & T_2 \\ T_3 & T_4 \end{bmatrix}$$

使(6.3.4)成立,即有

$$T_y = (-\eta U + V)T - T(\eta\sigma + P)$$

(6.3.5)

比较 η 的系数矩阵给出

$$UT + T\sigma = 0$$

(6.3.6a)

$$T_y = VT - TP$$

(6.3.6b)

则由(6.3.6a)给出 $T_2 = T_3 = 0$,由 $\det T = 1$,可以得到 $T_4 = \dfrac{1}{T_1}$,

代入(6.3.6b)给出

$$T = \begin{bmatrix} \mathrm{e}^{-y\frac{\beta}{\alpha}(1-\mathrm{e}^{-at})} & 0 \\ 0 & \mathrm{e}^{y\frac{\beta}{\alpha}(1-\mathrm{e}^{-at})} \end{bmatrix}$$

(6.3.7)

$$q = u e^{-\alpha t + 2y\frac{\beta}{\alpha}(1-e^{-\alpha t})} \qquad (6.3.8a)$$

$$r = v e^{-\alpha t - 2y\frac{\beta}{\alpha}(1-e^{-\alpha t})} \qquad (6.3.8b)$$

则有如下结论:

定理 6.1 设给定谱问题$(6.2.1)$与 $(6.2.2)$,其中 u, v 是 t, x 的足够光滑的函数,q, r 是 t, y 的足够光滑的函数,则通过一次自变量变换$(6.3.1)$和本征函数变换$(6.3.4)$,谱问题$(6.2.2)$化为谱问题 $(6.2.1)$,变换矩阵 T 可表示为$(6.3.7)$,而谱问题的位势之间满足关系式$(6.3.8)$.

§6.4 一阶非等谱方程族和等谱方程族之间转换算子

对于方程$(6.2.8)$,首先

$$e^{-\alpha t}(L' - 2b(t)) = e^{-\alpha t}\sigma\, e^{\alpha t}\partial_y - 2\frac{\beta}{\alpha}(1-e^{-\alpha t}) + 2e^{-\alpha t}\binom{u}{-v}\partial_y^{-1}(v e^{-\alpha t}, u e^{-\alpha t})$$

$$= \sigma\partial_y - 2\frac{\beta}{\alpha}(1-e^{-\alpha t}) + 2e^{-\alpha t}\binom{u}{-v}\partial_y^{-1}(v e^{-\alpha t}, u e^{-\alpha t})$$

$$= \begin{pmatrix} -\partial_y - 2\frac{\beta}{\alpha}(1-e^{-\alpha t}) & 0 \\ 0 & \partial_y + 2\frac{\beta}{\alpha}(1-e^{-\alpha t}) \end{pmatrix} +$$

$$2\binom{u e^{-\alpha t}}{-v e^{-\alpha t}}\partial^{-1}(v e^{-\alpha t}, u e^{-\alpha t})$$

$$= \sigma S\partial_y S^{-1} + 2S\binom{q}{-r}\partial_y^{-1}(r, q)S^{-1} \qquad (6.4.1)$$

其中

$$S = \begin{bmatrix} e^{-2\frac{\beta}{\alpha}(1-e^{-at})y} & 0 \\ 0 & e^{2\frac{\beta}{\alpha}(1-e^{-at})y} \end{bmatrix}$$

$$\alpha \begin{pmatrix} xu \\ xv \end{pmatrix}_x = \alpha \begin{pmatrix} yu \\ yv \end{pmatrix}_y \tag{6.4.2}$$

(6.3.8)可写成矩阵形式

$$S^{-1} \begin{bmatrix} ue^{-at} \\ ve^{-at} \end{bmatrix} = \begin{pmatrix} q \\ r \end{pmatrix} \tag{6.4.3}$$

进而有

$$S^{-1} \begin{bmatrix} yue^{-at} \\ yve^{-at} \end{bmatrix} = \begin{pmatrix} yq \\ yr \end{pmatrix}$$

即

$$\begin{bmatrix} yue^{-at} \\ yve^{-at} \end{bmatrix} = S \begin{pmatrix} yq \\ yr \end{pmatrix} \tag{6.4.4}$$

对(6.4.4)两端关于 y 求导,整理后得

$$\begin{bmatrix} yue^{-at} \\ yve^{-at} \end{bmatrix}_y = 2\frac{\beta}{\alpha}(1-e^{-at})\sigma S \begin{pmatrix} yq \\ yr \end{pmatrix} + S \begin{pmatrix} yq \\ yr \end{pmatrix}_y \tag{6.4.5}$$

将(6.4.5)代入(6.4.3),给出

$$\alpha \begin{pmatrix} yu \\ yv \end{pmatrix}_y = 2\beta(1-e^{-at})\sigma e^{at} S \begin{pmatrix} yq \\ yr \end{pmatrix} + \alpha e^{at} S \begin{pmatrix} yq \\ yr \end{pmatrix}_y \tag{6.4.6}$$

$$2\beta \begin{pmatrix} -xu \\ xv \end{pmatrix} = 2\beta \begin{bmatrix} -e^{-at}yu \\ e^{-at}yv \end{bmatrix} = 2\beta S \begin{pmatrix} -yq \\ yr \end{pmatrix} \tag{6.4.7}$$

另一方面,

由 $\begin{pmatrix} u \\ v \end{pmatrix} = e^{\alpha t} S \begin{pmatrix} q \\ r \end{pmatrix} = e^{\alpha t} \begin{pmatrix} e^{-2\frac{\beta}{\alpha}(1-e^{-\alpha t})y} & 0 \\ 0 & e^{2\frac{\beta}{\alpha}(1-e^{-\alpha t})y} \end{pmatrix} \begin{pmatrix} q(y(x,\ t),t) \\ r(y(x,\ t),t) \end{pmatrix}$

可得

$$\begin{pmatrix} u \\ v \end{pmatrix}_t = \alpha e^{\alpha t} S \begin{pmatrix} q \\ r \end{pmatrix} + 2\beta S \begin{pmatrix} -yq \\ yr \end{pmatrix} + 2\beta(1-e^{-\alpha t}) e^{\alpha t} S \begin{pmatrix} -q \\ r \end{pmatrix} + \alpha e^{\alpha t} x S \begin{pmatrix} q \\ r \end{pmatrix}_y +$$

$$e^{\alpha t} S \begin{pmatrix} q \\ r \end{pmatrix}_t = \alpha e^{\alpha t} S \begin{pmatrix} q \\ r \end{pmatrix} + 2\beta S \begin{pmatrix} -yq \\ yr \end{pmatrix} + 2\beta(1-e^{-\alpha t}) e^{\alpha t} S \begin{pmatrix} -q \\ r \end{pmatrix} +$$

$$\alpha y S \begin{pmatrix} q \\ r \end{pmatrix}_y + e^{\alpha t} S \begin{pmatrix} q \\ r \end{pmatrix}_t \qquad (6.4.8)$$

将(6.4.1)、(6.4.6)、(6.4.7)、(6.4.8)代入(6.2.8),有

$$\begin{pmatrix} u \\ v \end{pmatrix}_t - e^{-n\alpha t}(L'-b(t))^n \begin{pmatrix} -u \\ v \end{pmatrix} - \alpha \begin{pmatrix} xu \\ xv \end{pmatrix}_x - 2\beta \begin{pmatrix} -xu \\ xv \end{pmatrix}$$

$$= e^{\alpha t} S \left(\begin{pmatrix} q \\ r \end{pmatrix}_t - L^n \begin{pmatrix} -q \\ r \end{pmatrix} \right) \qquad (6.4.9)$$

定理 6.2 设给定谱问题(6.2.1)与 (6.2.2),其中 $u,\ v$ 是 $t,\ x$ 的足够光滑的函数而 $q,\ r$ 是 $t,\ y$ 的足够光滑的函数,如果位势$(u,\ v)$ 与$(q,\ r)$ 之间存在变换(6.3.8),则存在转换算子 $e^{\alpha t} S$

$$e^{\alpha t} S = \begin{pmatrix} e^{\alpha t - 2\frac{\beta}{\alpha}(1-e^{-\alpha t})y} & 0 \\ 0 & e^{\alpha t + 2\frac{\beta}{\alpha}(1-e^{-\alpha t})y} \end{pmatrix}$$

使得

$$\begin{pmatrix} u \\ v \end{pmatrix}_t - e^{-n\alpha t}(L'-b(t))^n \begin{pmatrix} -u \\ v \end{pmatrix} - \alpha \begin{pmatrix} xu \\ xv \end{pmatrix}_y - 2\beta \begin{pmatrix} -yu \\ yv \end{pmatrix}$$

$$= e^{\alpha t} S \left(\begin{pmatrix} q \\ r \end{pmatrix}_t - L^n \begin{pmatrix} -q \\ r \end{pmatrix} \right).$$

参 考 文 献

[1] C. S. Gardner, J. M. Greene, M. D. Kruskal, and R. M. Miura. Method for solving the Korteweg-devries equation. Phys. Rev. Lett. , 1967(19),1095 – 1097.

C. S. Gardner, J. M. Greene, M. D. Kruskal, and R. M. Miura. Korteweg-de Vires equation and generalizations. VI. methods for exact solution. Comm. Pure Appl. Math. , 1974(27), 97 – 133.

[2] R. M. Miura, Korteweg-de Vires equation and generalizations. I. existence of conservation laws and constants of motion. J. Math. Phys. , 1968(9), 1204 – 1209.

[3] L. D. Landau and E. M. Lifshitz. Quantum mechanics, nonrelativistic theory. Addison-Wesley, 1958.

[4] I. M. Gel'fand and B. M. Levitan. On the determination of a differential equation for its spectral function. Amer. Math. Soc. Transl. , Ser. 2,1(1955),253 – 304.

[5] P. D. Lax. Integrals of nonlinear equations of evolution and solitary waves. Comm. Pure Appl. Math. , 1968(21), 467 – 188.

[6] M. J. Ablowitz, D. J. Kaup and A. C. Newell. Nonlinear evolution equations of physical significance. Phys. Rev. Lett. , 1973(31), 125 – 127.

[7] M. J. Ablowitz, D. J. Kaup and A. C. Newell. The inverse scattering transform — Fourier analysis for nonlinear problems. Stud. Appl. Math. ,1974(53),249 – 315.

［ 8 ］ M. J. Ablowitz and R. Haberman. Resonantly coupled nonlinear evolution equations. J. Math. Phys. , 1975(16), 2301 - 2305.

［ 9 ］ V. E. Zakharov and A. B. Shabat. Exact theory of two-dimensional self-focusing and one-dimensional of waves in nonlinear media. Sov. Phys. JETP, 1972(34), 62 - 69.

［ 10 ］ M. Wadati. The modified Korteweg-de Vries equation. J. Phys. Soc. Jpn. , 1972(32),1681.

［ 11 ］ M. J. Ablowitz, D. J. Kaup, A. C. Newell and H. Segur. Method for solving the sine-Gordon equation. Phys. Rev. Lett. , 1973(30),1262 - 1264.

［ 12 ］ A. I. Nachman and M. J. Ablowitz. A multidimensional inverse scattering method. Stud. Appl. Math. , 1984(71), 243 - 250.

［ 13 ］ M. J. Ablowitz and A. I. Nachman. Multidimensional nonlinear evolution equations and inverse scattering. Phys. D，1986(18),223 - 241.

［ 14 ］ H. Flaschka. The Toda lattice，Ⅰ. Existence of integral. Phys. Rev. B，1974(9), 1924 - 1925.

［ 15 ］ M. J. Ablowitz and P. A. Clarkson，Solitons. Nonlinear Evolution Equation and Inverse Scattering. Cambridge university press,1991.

［ 16 ］ 李翊神,庄大蔚.两类非线性演化方程的等价.北京大学学报,1983(2),107—118.

［ 17 ］ 顾新身.相应的特征值问题有二重特征根的复 KdV 方程的一个求解实例.中国科学技术大学学报数学专辑,1983(5),82—89.

［ 18 ］ Y. B. Zeng，W. X. Ma and R. L. Lin. Integration of the soliton hierarchy with self-consistent sources. J. Math.

Phys., 2000 (41), 5453 - 5489.

[19] R. L. Lin, Y. B. Zeng and W. X. Ma. Solving the KdV hierarchy with self-consistent source by inverse scattering method. Phys. A, 2001(291), 287 - 298.

[20] T. K. ning, D. Y. chen and D. J. zhang. Soliton-like solutions for anonisospectral KdV hierarchy. Chaos, Solitons and Fractals, 2004(21), 395 - 311.

[21] T. K. ning, D. Y. chen and D. J. zhang. The exact solutions for the nonisospectral AKNS hierarchy through the inverse scattering transform. Physica A 2004 (339), 248 - 266.

[22] R. M. Miura. Bäcklund transforms, Vol. 515 in "Lecture Notes in Mathematics", pp40 - 68. Berlin, Springer, 1967.

[23] C. Rogers and W. F. Shadwick. Bäcklund transforms and their Applications. vol. 161 in "Mathematics in Science and Engineering." New York, Academic Press, 1982.

[24] 谷超豪等. 孤立子理论与应用. 杭州：浙江科学技术出版社, 1990.

[25] 谷超豪, 胡和生, 周子翔. 孤子理论中的达布变换及其几何应用. 上海, 上海科技出版社, 1999.

[26] R. Hirota. A new form of Bäcklund transformations and its relation to the inverse scattering problem. Progr. Theor. Phys., 1974(52), 1498 - 1512.

[27] 陈登远. 孤子引论. 北京：科学出版社, 2005.

[28] R. Hirota. Exact soliton of the KdV equation for multiple collision of solitons. Phys. Rev. Lett., 1971(27), 1192 - 1194.

[29] V. E. Zakharov and A. B. Shabat. Exact theory of two-dimensional self-focusing and one-dimensional self-modulation of waves in nonlinear media. ZhETF, 1971

(61),118, JETP, 1972(34), 62 - 69.

[30] R. Hirota. Exact envelope-soliton solution of a nonlinear wave equation. J. Math. Phys., 1973(14), 805 - 809.

[31] R. Hirota. Exact N-soliton solutions of the wave equation of long wave in shallow-water and in nonlinear lattice. J. Math. Phys., 1973(14), 810 - 814.

[32] R. Hirota. Exact three-soliton solution of the two-deminsional Sine-Gordon equtaion. J. Phys. Soc. Japan, 1973(35), 1566.

[33] R. Hirota. Exact solution to the equation discribing cylindrical solitons. Phys. Lett., 1979(71), 393 - 394.

[34] A. Nakamura and R. Hirota. Second Modified KdV equation and its exact Multi-Soliton Solution. J. Phys. Soc. Japan., 1980(48), 1755 - 1762.

[35] R. Hirota and M. Ito. A Direct Approach to Multi-periodic Wave Solitons to Nonlinear Evolution Equation. J. Phys. Soc. Japan, 1981(50), 338 - 342.

[36] R. Hirota. Exact N-soliton solution of Nonlinear Lumped self-daul Network equations. J. Phys. Soc. Japan, 1973(35), 289 - 294.

[37] R. Hirota. Nonlinear Partial Difference Equation. Ⅰ. A difference Analoguce of the KdV equtaion. J. Phys. Soc. Japan, 1977(43), 1424 - 1433.

[38] R. Hirota. Nonlinear Parial Difference Equation. Ⅱ. Discrete-time Toda Equation. J. Phys. Soc. Japan., 1977(43), 2074 - 2078.

[39] R. Hirota. Nonlinear Parial Difference Equation. Ⅲ. Discrete Sine-Gordon equation. J. Phys. Soc. Japan., 1977(43), 2079 - 2086.

[40] R. Hirota. Nonlinear partial difference equations. IV.

Bäcklund transformation for the discrete-time Toda equation. J. Phys. Soc. Japan. , 1977(45), 321 - 332.

[41] R. Hirota. Nonlinear partial difference equations. V. Nonlinear equations reducible to linear equations. J. Phys. Soc. Japan. , 1979(46), 312 - 319.

[42] R. Hirota, M. Ito and F. Kako. Two-dimensional Toda Lattice equation. Progr. Theor. Phys. Suppl. , 1988(94), 42 - 58.

[43] J. Satsuma. A wronskian representation of n-soliton solutions of nonlinear evolution equations. J. Phys Soc. Jpn. , 1979(46), 359 - 360.

[44] N. C. Freeman and J. J. C. Nimmo. Soliton solutions of the Korteweg-de Vries and Kadomtsev-Ketmiashvili equations:the wronskian technique. Phys. Lett. , A, 1983 (95), 1 - 3.

[45] J. J. C. Nimmo and N. C. freeman. A method of obtaining the n-soliton solution of the Boussinesq equation in terms of a Wronskian. Phys. Lett. , A, 1983(95), 4 - 6.

[46] R. Hirota. Solitons of the classical Boussinesq equation and the spherical Boussinesq equation:the wronskian technique. J. Phys. Soc. Japan, 1986(55), 2137 - 2150.

[47] I. Loris and R. Willox. Soliton solutions of Wronskian form to the nonlocal Boussinesq equation. J. Phys. Soc. Japan , 1996(65), 383 - 388.

[48] J. J. C. Nimmo and N. C. Freeman. The use of Bäcklundtransformations in obtaining N-soliton solutions in Wronskian form. J. Phys. A, 1984(17), 1415 - 1424.

[49] J. J. C. Nimmo. A bilinear Bäcklundtransformation for the nonlinear Schrödingerequation. Phys. Lett. , A, 1983 (99), 279 - 280.

[50] N. C. Freeman. Soliton Solutions of nonlinear evolution equations. IMA J. Appl. Math. , 1984(32), 125 - 145.

[51] J. J. C. Nimmo. Soliton solitons of three differential-difference equations in Wronskian form. Phys. Lett. , 1983(99), 281 - 286.

[52] R. Hirota, Y. Ohta and J. satsuma. Wronskian structures for soliton equations. Prog. Theo. Phys. Suppl. , 1988 (94), 59 - 72.

[53] R. Hirota, Y. Ohta and J. Satsuma. Solutions of Kadomtsev-Petviashvili equation and the two-dimensional Toda equations. J. Phys. Soc. Japan, 1988 (57), 1901 - 1904.

[54] Q. M. Liu. Solving formula of the two-dimensional Toda Lattice. J. Phys. A , 1989(22), 255 - 257.

[55] T. Tamizhmani, S. K. Vel and K. M. Tamizhmani. Wronskian and rational solutions of the differential-difference KP equation. J. Phys. A , 1998(31), 7627 - 7633.

[56] W. T. Han and Y. S. Li. Remarks an the solutions of the Kp equation. Phys. Lett. A, 2001(283),185 - 194.

[57] 韩文廷. 孤子方程特解的一种矩阵表示[学位论文]. 中国科技大学,2003.

[58] R. Hirota. Soliton solution to BKP equtaion. I. The Pfaffian Technique. J. Phys. Soc. Japan, 1989 (58), 2285 - 2296.

[59] Y. B. Zeng. Deriving N-soliton solutions via constrained flows. J. Phys. A, 2000(33), L115 - L120.

[60] Y. B. Zeng1 and H. H. Dai. Constructing N-soliton solutions for the mKdV equation through constrained flows. J. Phys. A, 2001(34), L657 - L663.

[61] D. Y. Chen, D. J. Zhang and S. F. Deng. The novel

multisoliton solution of the mKdV-sineGordon equation. J. Phys. Soc. Jpn., 2002(71),635 - 659.

[62] Z. B. Li and M. L. Wang. Travelling wave solutions to the two-dimensional KdV-Burgers equation. J. Phys. A: Math. Gen., 1993(26), 6027.

[63] 范恩贵,张鸿庆.非线性孤子方程的齐次平衡法.物理学报, 1998(47),353.

[64] W. Hereman, P. P. Banerjee, A. Korpel, G. Assanto, A. V. Immerzeele and A. Meerpoel. Exact solitary wave solutions of nonlinear evolution and wave equations using a direct algebraic method. J. Phys. A: Math. Gen., 1986 (19), 607.

[65] 李志斌,张善卿.非线性波方程准确孤立波解的符号计算.数学物理学报,1997(17),81.

[66] 刘式适,傅遵涛,刘式达,赵强.Jacobi 椭圆函数展开法及其在求解非线性波动方程中的应用.物理学报,2001(50),2068.

[67] M. R. Gupta. Exact inverse scattering solution of a nonlinear evolution equation in a nonuniform medium. Phys. Lett. A, 1979 (72), 420 - 422.

[68] W. L. Chan and K. S. Li. Nonpropagating solitons of the variable coefficient and nonisospectral Korteweg-de Vries equation. J. Math. Phys. 1989(30), 2521 - 2526.

[69] C. Tian and Y. J. Zhang, Nonlinear Physics (Reseach Reports in Physics) Berlin, Heidelberg: Springer-Verlag 1990, 35.

[70] C. Tian and Y. J. Zhang. Bäcklundtransformations for the isospectral and nonisospectral mKdV hierarchies. J. Phys. A: Math. Gen., 2867 - 2877.

[71] 李翊神.一类发展方程的谱变形.中国科学 A,1982 (25),911.

[72] W. X. Ma. An approach for constructing non-isospectral hierachies of evolution equations. J. Phys. A: Math. Gen. 1992 (25), L719 - L728.

[73] D. Y. Chen, H. W. Zhang. Lie algebraic structure for the AKNS system. J. Phys. A: Math. Gen., 1991(24), 377 - 383.

[74] D. Y. Chen, D. J. Zhang. Lie algebraic structure of (1+ 1)-dimensional Lax integrable systems. J. Math. Phys., 1996(37), 5524 - 5538.

[75] 张大军. Hereditariness and strong symmetry of the recursion operator for (1+1)-dimensional Lax integrable systems. 硕士学位论文,上海大学,1996.

[76] D. J. Zhang, D. Y. Chen. Hamilton structure of discrete soliton system. J. Phys. A: Math. Gen. 2002(35), 7225 - 7241.

[77] W. X. Ma. Lax representations and Lax operator algebras of isospectral and nonisospectral hierarchiesof evolution equations. J. Math. Phys., 1992(33), 2464 - 2476.

[78] W. X. Ma, B. Fuchssteiner. Algebraic structure of discrete zero curvature equations and master symmetry of discrete evolution equations. J. Math. Phys., 1999 (40), 2400 - 2418.

[79] A. C Newell. Solitons in Mathematics and Physics. University of Arizona, 1985.

[80] F. Calogero and A. Degasperis. The spectral Transform and Solitons. New York. North-Holland publishing company, Amsterdam. 1982.

[81] R. Hirota and J. Satsuma. A variety of nonlinear network equations generated from the Bäcklund transformation for the Toda lattice. Progr. Theoret. Phys. Suppl. 1976(59), 64 - 100.

[82] S. Sirianunpiboon, S. D. Howard and S. K. Roy. A note

on the Wronskian form of solutions of the KdV equation.
Phys. Lett. A, 1988 (134), 31 - 33.

[83] D.J. Zhang. Singular solutions in Casoratian form for two
differential-difference equations, Chaos, Solitons and
Fractals, 2005 (23), 1333 - 1350.

[84] N.C. Freeman and J.J.C. Nimmo. Soliton solutions of the
KdV and KP equations: the Wronskian technique. Phys.
Lett. A, 1983 (95), 1 - 3.

[85] R. Hirota. Exact soliton of the Sine-Gordon equtaion for
multiple collisions of solitons. J. Phys. Soc. Japan, 1972
(33), 1459 - 1463.

[86] D. J. Zhang. The N-soliton solutions for the modified
KdV equation with self-consistent sources. J. Phys. Soc.
Jpn., 2002(71), 2649 - 2656.

[87] D. J. Zhang, D. Y. Chen. The N-soliton solutions of the
sine-Gordon equation with self-consistent sources. Physica
A, 2003(321), 467 - 481.

[88] J. Wu, R. Keolian and I. Rudnick. Observation of a
nonpropagating hydrodynamic soliton. Phys. Rev. Lett.,
1984(52), 1421 - 1424.

[89] R. Wei , B. Wang, Y. Mao, X. Zheng and G. Miao.
Forced standing solitons and their unique propertties,
Proceedings of the Third Asia Pacific Physics Conference.
Singapore: WorldScientific, 1989, Vol.2, p.871.

[90] W. L. Chan and K. S. Li. Nonpropagating solitons of the
variable coefficient and nonisospectral Korteweg-de Vires
equation. J. Math. Phys., 1989(30), 2521 - 2526.

[91] V.E. Zakharov. On stocastization of one-dimensional
chains of nonlinear oscillations. Sov. Phys. JETP, 1974
(38),108 - 110.

［92］ M. J. Ablowitz and J. Satsum. Solitons and rational solutions of nonlinear evolution equations. J. Math. Phys. , 1978(19), 2180－2186.
A. I. Nachman and M. J. Ablowitz，A multidimensional inverse scattering for first order systems. Stud. Appl. Math. , 1984(71), 251－262.

［93］ M.J. Ablowitz and J. F. Ladik. Nonliear differential-difference equations. J. Math. phys. , 1975(16), 598－603. M.J. Ablowitz and J. F. Ladik. Nonliear differential-difference equations and Fourier analysis. J. Math. phys. , 1976(17), 1011－1018.

［94］ M. Wadati, · K. Konno and Y. H. Ichikawa. A generalization of the inverse scattering method. J. Phys. Soc. Jpn. , 1979(46), 1965－1966. H. Flaschka. On the Toda lattice，Ⅱ. Inverse scattering solution. Prog. Theo. Phys. , 1974(51), 703－706.

［95］ N. C. freeman. Soliton solutions of non-linear evolution equations. J. Appl. Math. , 1984(32), 125－145.

［96］ S. Kakei，N. Sasa and J. Satsuma. Bilinearization of a generalized derivative nonlinear Schrödingerequation. J. Phys. Soc. Japan , 1995(64), 1519－1523.

［97］ 张大军,邓淑芳. 孤子解的 Wronskian 表示. 上海大学学报，2002(8), 232—242.

［98］ Y. J. Zhang. Wronskian-type solutions for the vector k-constrained KP hierarchy. J. Phys. A, 1996（29）, 2617－2626.

［99］ F. Yuasa. Bäcklundtransformation of the two-dimensional Toda Lattice and Casorati's determinants. J. Phys. Soc. Japan，1987(56), 423－424.

［100］ J. J. C. Nimmo and N. C. Freeman，Rational solutions of

the Korteweg-de Vries equation in Wronskian form. Phys.
Lett, A,1983(96), 443 - 446.

[101] V. B. Matveev. Generalized Wronskain formula for
solutions of the KdV equations: first applications. Phys.
Lett. A, 1992(166), 205 - 208.

[102] Q. M. Liu. Double Wronskian solutions of the AKNS and
the classical Boussinesq hiersrchies. J. Phys. Soc. Japan,
1990(59), 3520 - 3527.

[103] Q. M. Liu, X. B. Hu and Y. Li. Rational solutions of the
classical Boussinesq hierarchy. J. Phys. A: Math Gen. ,
1990(23), 585 - 591.

[104] P. E. Hydon Conservation laws of partial difference
equations with two independent variables. J. Phys. A:
Math. Gen. , 2001(34), 10347 - 10355.

[105] A. S. Fokas Symmetries and integeability. Stud. Appl.
Math. , 1987(77), 253 - 299.

[106] B. Fuchssteiner, A. S. Fokas. Symplectic structures, their
Bäcklund transformations and hereditary symmetries.
Phys. D, 1981(4), 47 - 66.

[107] A. S Fokas, R. L. Anderson. On the use of isospectral
eigenvalue problems for obtaining hereditary symmetries
for Hamilton systems. J. Math. Phys. , 1982 (23), 1066 -
1073.

[108] M. Wadati, H. Sanuki and K. Konno. Realtionships
among inverse method, Bäcklund transformation and
infinite number of conservation laws. Prog. Theor. Phys.
1975 (53), 419 - 436.

[109] T. K. Ning, D. Y. Chen and D. J. Zhang. Exact Solutions
and conservation laws for a nonisospectral sine-Gordon
equation. Chaos, Solitons and Fractals, (Inpressed).

博士期间科研成果

已发表或录用的学术论文

[1] T. K. Ning, D. Y. Chen and D. J. Zhang. The exact solutions for the nonisospectral AKNS hierarchy through the inverse scattering transform. Phys. A, 2004 (339), 248 - 266.

[2] T. K. Ning, D. Y. Chen and D. J. Zhang. Soliton-like solutions for anonisospectral KdV hierarchy. Chaos, Solitons and Fractals, 2004(21), 395 - 311.

[3] T. K. Ning, D. Y. Chen and D. J. Zhang. Exact Solutions and conservation laws for a nonisospectral sine-Gordon equation. Chaos, Solitons and Fractals, (In pressed).

[4] 宁同科, 张大军, 陈登远. 一些非线性发展方程解的特征. 全国第四届孤立子与可积系统研讨会, 内蒙古师范大学, 2004 年 7 月 12 日—17 日.

[5] T. K. Ning, D. Y. Chen and D. J. Zhang. Solutions to τ-equations related to the AKNS spectral problem: Addendum to the exact solutions for thenonisospectral AKNS hierarchy through the inverse scattering transform. Phys. A (Accepted).

[6] 石教云, 宁同科, 张大军. 一个非等谱非线性 Schröodinger 方程的解及其无穷守恒律. 上海大学学报, 2004(10), 595—598.

[7] J. Y. Shi, T. K. Ning and D. J. Zhang. Soliton-like solutions of three non-isospectral equations. 上海大学学报 (英文版), (录用).

［ 8 ］　张大军,宁同科.无穷守恒律.上海大学学报(已提交修改稿).

［ 9 ］　宁同科,张大军,毕金钵.行列式的 Pfaffian 表示及其在孤子
　　　　解表示中的应用.(已投稿).

参加的科研课题

［ 1 ］　国家自然基金项目:孤子方程的新解与元胞自动机;

［ 2 ］　上海市教委项目:非等谱发展方程若干问题.

获奖情况

［ 1 ］　2003 年度上海大学 SMEG 二等奖学金;

［ 2 ］　2004 年度上海大学光华一等奖学金.

致　谢

首先感谢我的导师陈登远先生三年来对我学业上的悉心指导和生活上的无微关怀,先生对科学的执着追求,一丝不苟的治学态度,俭朴的生活作风,深深地影响着我,是我学习的典范,激励着我不断前进.三年来,我的每一个进步也同样凝结着张大军副教授的悉心帮助,他踏实的工作态度,朴实的人格都是我学习的榜样.在此谨向导师表示崇高的敬意和衷心的感谢.

感谢中国科技大学李翊神教授无私帮助和热情鼓励,感谢复旦大学数学研究所周子翔教授、范恩贵教授、上海交通大学物理系楼森岳教授和华东师范大学计算机系李志斌教授的有益帮助!

感谢马和平教授、盛万成教授、茅德康教授、秦成林教授、王远弟副教授以及其他关心我学习和生活的老师.同时感谢邓淑芳博士后、夏铁成博士后、张翼、孙业鹏、毕金钵博士的关心与帮助.

最后,特别感谢我的妻子靳慧敏女士和女儿宁方璞给予我理解与支持,感谢中国信达资产管理公司郑州办事处主任燕辉先生多年来的关心与支持.

谨以此文献给所有给予我指导、帮助、关心、支持的老师、同学、亲人和朋友们!

约 定 和 记 号

$\partial = \dfrac{\mathrm{d}}{\mathrm{d}x}$ 是微分算符，$\partial^{-1} = \displaystyle\int_{x}^{\infty}$ 是积分算符.

$\partial_{y} = \dfrac{\mathrm{d}}{\mathrm{d}y}$ 是微分算符，$\partial_{y}^{-1} = \displaystyle\int_{y}^{\infty}$ 是积分算符.

$P_{\mu} = \{u(x) \in \mathbf{R} \mid \displaystyle\int_{-\infty}^{\infty} (1 + \mid x \mid^{\mu}) \mid u(x) \mid \mathrm{d}x < \infty$ $(\mu = 0, 1, 2)\}$ 是一类实函数空间.

$W(u(x), v(x)) = u(x)v_{x}(x) - u_{x}(x)v(x)$ 表示函数 $u(x)$ 和 $v(x)$ 的 Wronskian 行列式；用 $<\cdot, \cdot>$ 表示 Hilbert 空间的内积 $\displaystyle\int_{-\infty}^{\infty} \cdot \mathrm{d}x$.

$\sigma = \begin{pmatrix} -1 & 0 \\ 0 & 1 \end{pmatrix}.$

$(\cdot)^{*}$ 表示 $(\cdot)$ 的共轭.